Secondary Ion Mass
Spectroscopy of Solid Surfaces

Secondary Ion Mass
Spectroscopy of Solid Surfaces

V. Cherepin

*Institute of Metal Physics, Academy of Sciences of the
Ukrainian SSR, USSR*

CRC Press
Taylor & Francis Group
Boca Raton London New York

CRC Press is an imprint of the
Taylor & Francis Group, an **informa** business

Contents

Preface

Characterization of solid surfaces is one of the major problems in modern science. Since materials interact with each other and the environment through the surface, it is the outer surface layer that often determines the behaviour of the bulk. In recent years a drastic increase in the number of publications can be observed in this progressing field. The progress achieved in solid surface studies is due mainly to the applications of modern physical methods. In this book entitled Secondary Ion Mass Spectroscopy (SIMS) a most important physical method of surface analysis is described.

A companion volume to this work, a book covering an equally important method of surface anlaysis, *X-ray Photoelectron Spectroscopy (XPS)* by V.I. Nefedov is available from the same publisher.

SIMS can be applied to all elements and allows the quantitative analysis of solid surfaces, including monolayers. The method is widely used in studies of adsorption, oxidation, corrosion, catalysis, diffusion and in the characterization of thin films and coatings. In addition, SIMS can be used in the study of isotopic composition and has proven to be very effective in investigations of small impurities. All these items are considered in detail.

Researchers, both physicists and chemists, using SIMS are confronted with numerous publications on the subject, scattered over many scientific journals. Therefore, in this monograph an attempt is made to summarize the available material in a systematic way. The monograph has two objectives. One is to meet the interests of both experienced workers as well as those just starting their activities in the field. Accordingly, Chapters 1 and 2 are devoted to the systematic treatment of the physical backgrounds of SIMS. Special attention has been paid to the consideration of important data in surface studies. Extensive data, necessary for the quantitative analysis in particular, are presented. Secondly, the interests are taken into account of those researchers in scientific and industrial organizations who wish to get acquainted with the scope of the method and the results obtained in a particular application. This purpose is met in Chapters 3–5 where the main results are summarized concerning catalysis, oxidation, corrosion, adhesion, implant profiles, surface compounds and other important fields.

The original Russian text has been substantially revised and amended during the preparation of this English translation. New information appearing in the literature upto the end of 1985 has been included. Each chapter is supplied with an extensive list of references.

The author wishes to express his deep appreciation to L.G. Ryaboshapka, I.V. Pasechnik, V.I. Is'yanov, and E.E. Dunaeva for their assistance in the preparation of the English manuscript.

V.T. Cherepin
Kiev, February 1986

Chapter 1
Physics of secondary ion emission

1.1. Parameters of secondary ion emission

Energetic ion bombardment of a solid surface causes sample atoms to shift from their original states as positive and negative ions known as secondary ions. Highly sensitive mass spectrometric measurement of the secondary ion quantity creates the basis of methods developed for the constitutional analysis of solid surfaces and the bulk. The most important features of the method of secondary ion mass spectrometry (SIMS) are a very low sensitivity limit for majority of elements (less than 10^{-4} monolayer), the possibility of measuring the concentration profiles of implanted impurity with a depth resolution better than 5 nm, spatial resolution in the micrometer range, the possibility of isotope analysis, and identification of all the elements and isotopes starting from hydrogen.

SIMS is now a well-established method with special instrumentation and sophisticated techniques for analysis of various objects. Many physical aspects of secondary ion emission have been studied in detail, but at the same time there are many problems still to be solved. The method is widely used for investigations not only in fundamental laboratories, but also in industry, and further extension of its application in various fields of science and technology should be expected.

Since the mid-1960s, interest in SIMS as a method for the analysis of solid surfaces and bulk has been continually growing, as may be judged from the number of publications on this subject over the past 10–12 years, as well as from comprehensive reviews and monographs which give some survey of completed work [1–10].

In order to describe the process of secondary ion emission (SIE) and to establish the relationship between the quantity of ions and physical and chemical properties of surfaces being bombarded, the following coefficients are used:
- the SIE coefficient $K_i^+ = N_i^+/N_0$, where N_i^+ is the number of positive secondary ions with a definite charge to mass ratio and N_0 is the number of primary ions;
- the secondary ion yield for multicomponent targets, $\gamma_i^+ = K_i^+/C_i$ (C_i is the concentration of the ith component);

1

● the sputtering coefficient $S = N/N_0$, where N is the total number of sputtered particles, i.e. the sum of neutral (N^0) and ionized (N_i^+) particles;
● the ionization efficiency (or ionization probability) $R_i^+ = K_i^+/S$, which characterizes the fraction of ions in the total number of sputtered particles (R_i^+ varies from 0 to 1);
● the degree of ionization $\alpha_i^+ = N_i^+/N^0$, i.e. the number of sputtered ions related to the number of atoms sputtered in the neutral state (α_i^+ may have the values from 0 to ∞). At $R_i^+ \ll 1$, $\alpha_i^+ = R_i^+$.

Since the absolute values of the above parameters cannot be easily determined, the SIE processes are sometimes characterized by relative values of these coefficients. For instance, the relative coefficient $K_{i\,rel}^+ = K_i^+/K_{ref}^+$, where K_{ref}^+ is the coefficient of the SIE for a reference sample.

The absolute K_i^+ value may be determined by measuring an integral coefficient of secondary ion emission, $K : K = \sum_{(i)} N_i/N_0$, where $\sum_{(i)} N_i$ is the total number of all the secondary ions ejected from the surface by the flux of primary ions N_0.

Knowledge of the absolute K^+ value makes it possible to define the absolute K_i^+ values by multiplying K by the magnitude of a relative intensity corresponding to the mass spectral line in the mass spectrum of the secondary ion emission (provided the angular distributions of all secondary ions are the same).

In order to understand the mechanism of the emission and to design the secondary ion optics in analytical instruments, the angular and energy distributions of secondary ions must be known, as must also the effect of primary ions energy and current, target material etc. Typical laboratory installations for the solution of such problems are described in [11–18].

1.2. Effect of the primary ion energy and current on SIE

Secondary ion emission begins when the primary ion energy exceeds some threshold level [18]. This level amounts to 30–80 eV and is much higher than the energies for sputtering of neutral particles [9].

With increasing primary ion energy intensity, the secondary ion emission becomes, as a rule, higher. Thus in [12] the K^+ (E_0) dependences have been obtained for the ions Cu^+, Ni^+, Fe^+, Si^+, Al^+, and Mg^+ ejected from the corresponding targets by Ar^+ ions. For all the targets studied in [12] (with the exception of Al) the SIE coefficient is the function of energy E_0. The relationships observed are close to the $S(E_0)$ characteristics for the neutral component of sputtering [19].

More detailed study of this effect with broader selection of targets (14 elements) and under better experimental conditions was performed in [20]. The secondary ion intensities at the maximum of energy distribution as a function of the primary ion energy for Be, Al, Si, V, Nb, Mo, W, Pt, and Au, corrected for natural abundance, are shown in Fig. 1.1 [20]. As one can see from Fig. 1.1 the secondary ion intensity increases with increasing energy for all the elements investigated. However, there are considerable differences in yield variation between some of the elements. Comparison at either 5 or 15

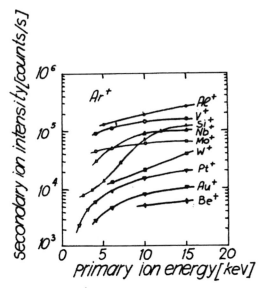

Figure 1.1 Secondary ion intensity Y^+ versus primary ion energy (normalized to a beam current of $1\mu A$) [20].

keV, for example, shows a 30% yield change in the case of Mo, whereas a change of a factor of 10 is observed with Si.

The primary ion current density is an important experimental parameter, which affects the accuracy of secondary ions measurement. In the general case the number N_i of secondary ions of the ith component may be given by the sum

$$N_i = N_i' + N_i'' \tag{1.1}$$

where N_i' and N_i'' are the number of secondary ions of the surface and the number of secondary ions of bulk origin, respectively. They are the function of the primary ion beam current density.

In [21], as a result of investigation of the adsorption dynamic equilibrium between gases and metal surface, the following expression has been deduced, which relates N_i' to the number N_0 of the primary ions per unit surface area:

$$N_i' = ApN_0 (B + CP + DN_0) \tag{1.2}$$

Here p is the gas pressure in the target region; A, B, C, D are constants. It follows from this expression that with growth of N_0 the number of ions of surface origin tends to a constant value Ap/D. This value will be reached at a smaller N_0 value, the lower the gas pressure in the vacuum chamber where the target is placed. At a sufficient large value of N_0, the degree of the metal surface covered by absorbed gas particles and by molecules of chemical compounds formed by gas particles and metal atoms becomes neglegibly small. Then the relation between N_i and N_0 may be written as

$$N_i'' = R_i^+ n_i N_0 \, \gamma \, (\theta) \tag{1.3}$$

Here N_i is the bulk concentration of the ith component in the sample under study; R_i^+ is the ionization eficiency; γ (θ) is the factor that accounts for the screening action of particles present at the surface and γ $(\theta) \rightarrow$ I for large N_0. Substituting the N_i' and N_i'' values from (1.2) and (1.3) in Equation (1.1), one obtains the general expression for dependence of N_i on N_0:

$$N_i = ApN_0/(B + C + DN_o) \tag{1.4}$$

It follows from the above relations that for definite N_0 values the N_i dependenc of N on N_0 becomes linear and may be written as

$$N_i = a + bN_0 \tag{1.5}$$

where $C = Ap/D$ and $B = R_i^+ n_i$ is a constant. In this case the tangent of the straight line slope is proportional to the bulk concentration of the ith component.

If the residual gas pressure and composition as well as the primary beam density are such that the inequality $N_i'' \gg N_i'$ is satisfied, the contribution from the ions of surface origin (N_i') may be ignored. In this case Equation (1.5) has the form:

$$N_i = N_i' = bN_0 \tag{1.6}$$

and the number of ions N_i emitted per unit area of the target surface may be taken as the quantity that characterizes the concentration of the ith component.

When the current density is increased to 3–5 mA cm^{-2}, the anomalous growth of the secondary ion emission may be observed. This effect is due to ionization processes in the gaseous phase because of a direct interaction between primary ions and sputtered neutral atoms [22, 23]. Such an interaction process may be easily revealed in materials with a high sputtering coefficient (e.g., gold, tin, zincum).

1.3. Effect of the target temperature on the SIE coefficient

The dependence of SIE coefficients or temperature was systematically studied by Y. M. Fogel [24]. On the basis of his work, techniques were developed for the application of SIE in investigations of gas–metal surface interaction [24, 25].

Since the secondary ion current I_i^+ is proportional to the SIE coefficient K_i^+, to the surface concentration of the particles under study n_i, and to the primary current I_0 (i.e. $I_i^+ = I_0 K_i^+ n_i$), the information about the processes to be investigated is obtained by plotting the dependence of I_i^+ on the target temperature and on the pressure of the gas, which interacts with the target. Therefore for a correct determination of n_i, a constant value of K_i^+ is required. The temperature dependence of K_i^+ might be revealed if I_i^+ and n_i could be measured simultaneously, but there are great difficulties in measuring the latter quantity.

Hence it has been a general custom to judge the K_i^+ (T) dependence from indirect data based on the study of temperature dependence of secondary ion currents. At low enough target temperatures the I_i^+ value decreases, as a rule, with temperature rise. Only at high enough temperatures does the number of sputtered ions remains constant [23–26]. This case corresponds to the sputtering of ions from the metal lattice. Below this temperature the metal surface is covered by oxide, from which metal ions are also ejected. The intensity of the ion beam Me$^+$ begins then to change with temperature in accordance with variations in the coverage of the metal surface by oxides, MeO. This may be seen from comparison of $I_i^+(T)$ curves for Me$^+$ and MeO$^+$ ions. If K_i^+ was temperature dependent, the $K_i^+(T)$ function would be of a monotonic character. However in many cases $I_i^+(T)$ curves have a non-monotonic run, due perhaps to variation in the surface concentration of particles of adsorbed gases and molecules of compounds formed at the surface [2, 24].

Another source of secondary ion yield variation is a possible change in the target crystal lattice properties due to phase transformations. Usually these effects are masked by more essential surface effects, but under appropriate experimental conditions they may be revealed and observed in correlation with variation of sputtering yield [27]. This is the case for phase transformations of the first and the second type, and this phenonemnon is very interesting from the physical point of view. However, the effects of phase transformations are not so appreciable as to create difficulties in performing compositional analysis.

1.4. Angular dependence of secondary ion emission

1.4.1. Single Crystals

Yurasova [17] was the first to demonstrate that at the normal bombardment of Cu(001) by Ar$^+$ and Ne$^+$ ions, distinct anisotropy may be observed in the angular distribution of secondary ions. Coefficients of anisotropy, which are determined as the ratio of differences in ion current maxima and minima to the current maxima, turned out to be highest for ions of both Ne$^+$ and Ar$^+$ (4.2 keV) at incident angles of $\theta = 40$ and $50°$; these angles correspond to the direction of a close packing <110>.

With growth of the ion energy from 4 to 6 keV no changes were observed in the secondary ion distribution over the azimuthal angle. Increasing the energy to 20 keV leads to narrower maxima in the angular distribution.

The change of the angle of incidence does not cause any significant variations in angular distribution, but at oblique incidence emission in the direction <100> can be also observed.

One of the main conclusions made in the cited works is coincidence of pictures of secondary ion distributions over the emission angles for normal and oblique incidence of the primary ion beam and the distribution of neutral sputtered particles under the same conditions. This feature was supported later in other works [28].

Anisotropy of the SIE coefficient K_i^+, as a function of azimuthal angle α, has been studied for a target rotating around an axis normal to its surface [29, 30]. Mo(001), (101), and Nb (100) single crystals were studied. The targets were bombarded by 8 keV Ar$^+$ ions at an incident angle of 60°, the current density being 1.5 mA cm^{-2}. The azimuthal angle α was varied from 0 to 360°.

Figure 1.2 shows the ^{98}Mo$^+$ ion current as a function of the target azimuthal angle at the bombardment of (001), (101), and (111) faces. It may be seen that the angular dependence has a periodic character, the number of ion current minima and their angular position being strongly dependent on the indexes {hkl} of the faces under study. Thus for example at the bombardment of the (001) face the minima are observed every 45 and 90°, while for the (111) face they occur every 30 and 60°. For (101) the current minima correspond to the angles, 0, 35, 75, 105, 145, 180, 215, 255, 285 and 325°.

These data may be understood in terms of the channelling theory developed by Lindhard [31] with the assumptions proposed by Underdelinden [32] for description of the orientation dependence of the sputtering coefficient: (1) the chanelled part of the beam does not contribute to sputtering; (2) the action of disordered beam is equivalent to the sputtering caused by the beam of the same intensity on the polycrystalline target. If we assume that the degree of ionization is independent on the angular position of a single crystal [33], the chanelled part of the beam may be considered as practically negligible not only for sputtering, but also for SIE, since there is no source of particles suitable for ionization.

Hence the ion current minima are connected with the primary ion channelling, while the maxima are connected with the transition to the 'disordered' orientation of the crystal. It has been shown [33–36] that the bombardment of a crystal in 'transparent' directions leads to sharp drop in the sputtering because of the lower probability of surface atoms being ejected since the collision cascade is developed in deeper layers. This circumstance is the main

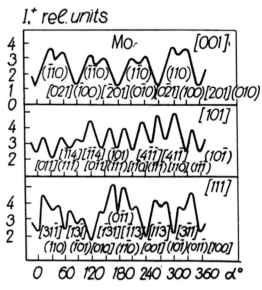

Figure 1.2 ^{98}Mo$^+$ secondary ion current versus rotation angle of Mo single crystal around the axes [001], [101], [111].

reason for the lower SIE coefficients when observed single crystal is bombarded in the direction of channelling.

It is worth mentioning that the primary beam channelling also results in lower-intensity photon emission [37, 38] and a lower coefficient of ion–electron emission [39]. This correlation shows that secondary ion intensity becomes weaker because of the decreasing number of excited target atoms, which are the source of ions, photons, and electrons.

With the discovery of the SIE orientation effects it became necessary to solve practical problems of how to suppress these effects during elemental or in-depth analysis of single crystals. The simplest ways of doing this are based on oxygen flooding of the sample or on defocusing the primary ion beam. For many single crystals these procedures ensure complete suppression of the angular distribution anisotropy. More complex ways proposed in [34, 35] consist of target bombardment by several ion beams from different directions, or in the use of heavy primary ions (W^+, $Al\,F^+$, $W\,F_5^+$).

1.4.2. Polycrystalline Metals

The dependence of the SIE coefficient, K_i^+, on the angle of incidence, θ, of bombarding ions, has been studied [12] for Al and Cu targets (Table 1.1). As may be seen from the table, K_i^+ increases for larger values of θ.

Angular distributions for secondary ions at normal and oblique incidence have been obtained [12, 40]. The targets were Mg, Al, Ti, Fe, Ni, and C. Ions of noble gases were used for bombardment. These distributions, for monoatomic singly charged Cu and Al ions at normal incidence, follow approximately the Lambert law, i.e. the intensity of emission is proportional to the cosine. However in directions close to the normal this law is not obeyed: distribution becomes more 'flat' or even 'concave' (e.g., for Mg). Figure 1.3 gives some typical angular distributions for secondary ions of various metals. Using Kr^+, Xe^+, and Ne^+ ions instead of Ar^+ does not bring any changes in the angular distributions. However, in the case of He^+ ions the angular distribution of Al^+ ions follows in fact the cosine law; angular distribution of C^+ ions becomes 'elongated' as should be expected from this law.

When the ratio of the target atom mass to the bombarding ion mass is increased, the angular distribution becomes still more elongated in the direction normal to the surface. This fact may be explained by the increasing role of the primary ion back-reflected onto the target atoms in a subsurface layer. As a result, a larger number of particles are ejected in the normal direction.

Table 1.1
K_i^+ as a function of angle of incidence for primary Ar^+ ions ($E_p = 8\,keV$) [12]

Target	K_i^+				
	0°	15°	30°	45°	60°
Al	2.5×10^{-2}	2.5×10^{-2}	—	3.5×10^{-2}	4.6×10^{-2}
Cu	6×10^{-4}	—	7×10^{-4}	1.3×10^{-3}	2.4×10^{-3}

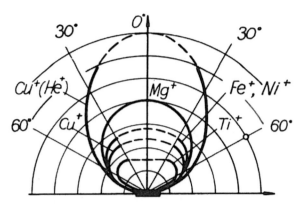

Figure 1.3 Angular distribution of secondary ions Mg^+, Ti^+, $Fe^+ Ni^+$, and Cu^+ at the bombardment of corresponding targets with 8 keV ions.

1.4.3. Clean Surfaces

The observations described above are useful from a practical point of view for designing the secondary ion collection optics to provide high measurement efficiency, but they give little help in the study of an actual clean surface. The primary beam used in traditional experiments is too crude both in energy and density of current—any original surface structure is completely destroyed at the very initial stages of bombardment. New trends in this field are based on angular and energy-resolved measurement of secondary ion emission in UHV conditions with low energy density ion bombardment [41–43]. Precise control of all the parameters allows study of clean and adsorbate-covered single-crystal surfaces under conditions that correspond to those modelled by methods of molecular dynamics calculations [44].

The information most sensitive to surface structure is contained in azimuthal angular spectra. These spectra, obtained at large polar angles, are strongly influenced by image force, which acts so as to bend the trajectories of charged particles, originally ejected at smaller polar angles, into the detector. In Fig. 1.4 the angular spectra obtained for 3 ± 3 eV Ni ions ejected from Ni(001) c $(2 \times 2) - CO$ are shown at polar angles $\theta = 30°$, $45°$, $60°$, and $70°$ (from [44]). Predicted neutral and image-force-corrected distributions are also shown, derived assuming an E_{image} value of 3.6 eV. Although the magnitude of the measured anisotropy is slightly smaller than that calculated at $45°$ and $30°$, the level of agreement is quite remarkable. These types of angle-resolved experiments should be valuable aids in the analysis of unknown surface structures and in understanding the ionization mechanism.

1.5. Secondary ion energy distribution

Energy spectra are the subject of many works devoted to the study of secondary ion emission. Comprehensive reviews of these works are given in [1–10]. A detailed study of secondary ion spectra provides a wealth of information about the ionization and sputtering processes. The shape of the

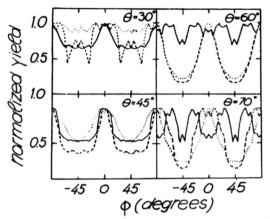

Figure 1.4 Azimuthal angle distributions at various polar angles for Ni ejected from Ni (001) c (2×2) – CO. Only those Ni particles with an energy of 3 ± 3 eV are detected. The value of φ = 0° corresponds to <100>, while φ = ± 45° corresponds to <110>.

energy spectrum can give clues about the dominant mechanism of ionization. The most probable energy of the distribution can be related to the binding energy of the surface, and the shape of the sputtered cluster ion energy distributions can be compared with theoretical predictions [45].

The secondary ion energy distribution may be measured well enough by means of a simple retarding field formed by hemispherical or plane grids. Better accuracy can be obtained with electrostatic deflection-type analyzers: cylindrical [43, 46, 47], toroidal, hemispherical [48], with cylindrical mirrors [49], and also with electrostatic mirrors [50]. Since a fine structure is not observed in the dN/dE distributions, the use of a high-resolution energy analyzer is unreasonable, and a modest 2–3 eV resolution is quite sufficient.

In an early investigation [40] of the energy distribution of Al^+ and Cu^+ secondary ions ejected from Al and Cu targets bombarded by 8 keV Ar^+ ions at normal incidence, it was shown that the mean energy $\bar{E}$ increases with increasing ejection angle. The following values were found for mean energies at the collection angle 30°: $E = 50$ eV for Al^+ and 160 eV for Cu^+. The secondary ion energy distribution may be measured well enough by means of a simple retarding field with an accuracy ± 20%. Typical energy spectra for Al^+ and Cu^+ obtained by differentiation of retarding field curves are shown in Fig. 1.5.

It has been found for Cu single crystals [15, 17] that the energy spectrum for Cu^+ ions ejected in the direction of closest packing (110) is narrower than for ions ejected in intermediate directions. The difference in spectra become less pronounced with increasing secondary ion energy. Crystallographic effect is more evident when a single Cu crystal is bombarded by Ar^+ ions with higher energy (10 keV) [51]. For example, the yield of ions with the most probable energy in the direction <110> is about 2.5 times higher than the yield in an intermediate direction.

In the first work in which a large number of objects (15 polycrystalline

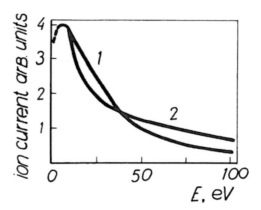

Figure 1.5 Energy spectra of Al^+(1) and Cu^+(2) secondary ions sputtered from corresponding targets by 8 keV Ar^+ ions.

metals) have been studied under the same experimental conditions [52] the following main observations were made:

(1) The energy curves are peaked, with a maximum at 10–100 eV;

(2) The shape of a low-energy part of the curve may be approximated by the function $N^+(E) \sim \exp(-mE)$, where $0.01 < M0.10$. The shape of a high-energy part of the curve may be described by the function $N^+(E) \sim E^{-n}$, where $1 < n < 2$. For Ar^+ ions the fitting parameter $n = 1.4$.

(3) The parameters of the energy distributions have certain periodical variation depending on the atomic number Z of the target.

An important characteristic of an energy spectrum is the most probable secondary ion energy. The method to define this value has been developed in [53], where it has been shown that for 8-keV Ar^+ ion bombarded Cu and Ni targets the most probable energy for Cu^+ ions is 4.5 eV, and for Ni^+ ions 3.5 eV. For energy higher than 30 eV, Cu^+ and Al^+ ions have practically identical energy distributions. The energy distributions for Cu_2^+, Cu_3^+, and Ni_2^+ polyatomic ions were also obtained in the same work and it has been found that the energy dispersion for complex ions is much lower than for monoatomic ions. The most probable energy for cluster ions is generally lower and is of the order of several eV. The first observations lead to the conclusion that the most probable energy of ions is related to the binding energy of the atom being sputtered. This line of investigation has been successfully developed in [54–56].

Several works were specially devoted to the effect of oxygen flooding on the shape of the secondary ion energy spectrum [57, 58]. It has been shown that O_2 absorption causes the most probable energy to be lowered by about 1 eV. The number of fast ions is not changed at oxidation, but the number of slow ions becomes essentially larger. This implies that the energy spectrum of ions ejected from oxides is narrower than that of ions from pure metals.

Secondary ion energy distributions for 8-keV Ar^+ ion bombarded polycrystalline metals selected from different groups of the Periodic Table were studied in detail in [56]. Analysis of the experimental data have shown that the highest concentration of slow ions among all the sputtered particles is observed for gold. The largest number of fast ions refers to the energy spectrum of V. Secondary ions with energy above 100 eV were not detected in Au spectra. It may be concluded from the comparison of energy distributions for various elements that the number of ions in a given energy range may strongly differ for different metals due to different shape of $N^+(E)$ curves. Elements differ by the values of secondary ion average energy $\bar{E}$, by FWHM of energy distributions $E_{0.5}$ and by the most probable (maximum) energies E_{max}. Average energies were found from the relation

$$E = \int_{E_0}^{E_{max}} EN^+(E)\, E/N^+(E)\mathrm{d}E \qquad (1.7)$$

It has been shown that Au and Co have the 'sharpest' distributions, with minimum values of average energies. V and Mn have the highest average energy values. Average and most probable ion energies are decreased with the filling of $3d$ shells (V, Mn, Co). For the elements of one group (V, Nb, Ta) a tendency to lower $\bar{E}$ and E_{max} is observed for higher atomic number. Similar periodic character was found earlier for the average energy of neutral sputtered atoms $\bar{E}_n$ [20].

A comprehensive study of secondary ion energy spectra obtained from pure elemental samples under conditions appropriate for analysis using SIMS was carried out by Rudat and Morrison [45]. For their experiments, three basic conditions were used: residual vacuum or low argon pressure backfill, high oxygen pressure backfill, or high nitrogen pressure backfill. Thirty-one pure elements and two compounds were sputtered by 5.5 keV O_2^+ ions with current density $I - 2$ mA cm^{-2} for cleaning and ~ 100 μA cm^{-2} for the spectra determinations.

The atomic ion energy spectra from all three conditions can be classified according to their appearance in the nine general types shown in Figs 1.6 and 1.7. The classification of the spectra for particular elements is indicated in Table 1.2. and may be used for proper settings of energy analysers in SIMS.

It should be mentioned that although the sublimation energy does not correlate well with the most probable energy, average energies of the atomic ions correlate with the sublimation energy of the metals, as do the average energies of the neutral distributions (Table 1.2) [45] This type of correlation is found for all three ambient gas conditions studied, with the best correlation occuring for spectra in the presence of oxygen gas. In general, the energy spectra of ions sputtered from pure metals by a reactive gas ion beam are very similar to those found from noble gas ion bombardment.

1.6. Effect of primary ions and target material on secondary ion yield

Secondary ion yield for a pure monoelement target is equal to the coefficient of secondary ion emission K^+, and this coefficient is one of the most important quantitative parameters in SIMS. It must be known for setting relative sensitivity factors in any type of analysis and detection limits for

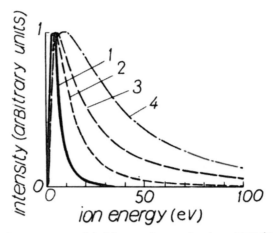

Figure 1.6 Typical energy spectra of positive atomic secondary ions: (1) Na^+/c; (2) Mg^+; (3) Ti^+; (4) Cr^+ (all in residual vacuum/low argon pressure) [45].

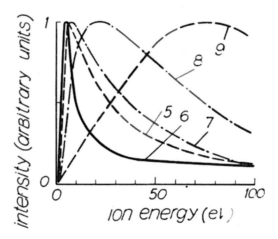

Figure 1.7 Typical energy spectra of positive atomic secondary ions: (5) Co^+; (6) Cu^+; (7) Hf^+; (8) V^+ (all residual vacuum/low argon pressure); (9) V^+ in 5×10^{-6} Torr O_2. [45].

different elements. Ionization efficiency is also determined from K^+ provided the sputtering yield S is known for the same target and same experimental conditions. The absolute value of K^+ is very difficult to measure and usually relative coefficients or useful output yields are compared for different elements. One must be careful comparing data obtained by different authors since secondary ion mass spectra and currents depend strongly on experimental conditions, first of all, on residual gas pressure and composition, on the state and composition of the target surface, on composition and purity of primary ion beam.

First systematic study of the ion yield for various elemental targets bombarded by inert gas ions was carried out in [12, 59]. Similar measurements were continued later by other authors so that a vast amount of

Table 1.2
Classification of positive atomic ion energy spectra by appearance

Type	Vacuum/low Ar pressure	High O_2 pressure	High N_2 pressure
1	H, Li, Na, K (impurity ions)	H, Li, Na, K (impurity ions)	H, Li, Na, K (impurity ions)
2	Mg, Al, P, Ga, Ge, In, Sb, Te, Pb, Bi (O from some samples)	Mg, Al, Ga, Ge, Te, Pb, Bi	Mg, Al, Ga, Ge, Sb, Te, Pb, Bi
3	B, Si, Ti, V, Mn, As (O from some samples)	B, Si, Ti, V, Cr, Mn, Fe, Co, Ni, Mo, Sb, Ta, W	B, Mn, Ta
4	C, Cr, Fe, Zn, Hf, Ta, W	C, Zn	C, Cr, Fe, Zn, Hf, W
5	Co, Ni, Mo		Co, Ni, Mo
6	Cu, Ru, Pd, Ag, Sn	Cu, Ru, Pd, Ag, Sn	Cu, Ru, Pd, Ag, Sn
7	Zr, Nb	Zr, Nb, Hf	Zr, Nb
8	U		U
9		U	

experimental data has been accumulated. Analysis of these results has shown that wide scattering of data on K^+ is observed for the same element measured in different laboratories [23] and variation of emission may differ by more than one order of magnitude. The strongest scattering is observed for elements with high oxygen affinity thus indicating an important role played by small uncontrollable variations in pressure and composition of residual atmosphere. However, despite this scattering of data definite conclusions on the mode of K^+ variation from element to element may be drawn. Exact regularities have been established as a result of systematic study of a large number of materials (45 pure metals) under well-controlled and reproducible experimental conditions [60–62].

Pure polycrystalline metal targets were bombarded by 8 keV Ar^+, He^+ and O_2^+ ions with current density $1.2 - 2.0$ mA cm^{-2} to ensure the yield of ions of bulk origin. The spectrum of sputtered ions was recorded after stable levels of secondary currents had been reached. Peak currents for all isotopes were normalized for the constant primary current and then summed, if necessary, for polyisotopic elements. For a constant primary current the resulting secondary currents are proportional to K^+. Actual currents measured with a magnetic sector type mass spectrometer are presented in Table 1.3 for 45 metals. Elements are arranged in the order of decreasing secondary currents for Ar^+ primary ions. Some important conclusions can be drawn from these data. First of all, it must be pointed out that secondary ion emission depends both on the properties of target material and the nature of primary species, and may differ in intensity by three orders of magnitude for Ar^+ bombardment (e.g., Lu and Cd), by three and a half orders of magnitude for He^+ ions (Al and Pt), and by four orders for O_2^+ (Al and Pt). In this respect SIE is the most structure sensitive phenomenon as compared to such processes as sputtering and ion–electron emission.

Some idea about the factors governing the processes of secondary ion

Table 1.3
Secondary ion currents for elements

Element	$I^+, 10^{-13}A$			Element	$I^+, 10^{-13}A$		
	Ar^+	He^+	O_2^+		Ar^+	He^+	O_2^+
Lu	55.3	9.33	95.6	Cr	1.67	6.18	138
Mg	24.6	18.8	208	Fe	1.66	1.90	49.7
Sc	22.2	115	1000	Rh	1.47	1.67	107.2
Tb	19.6	12.0	107.5	Ru	1.46	2.26	94.2
Al	15.3	212	1835	Sm	1.34	8.25	115
Nb	15.2	5.20	100	Bi	1125	0.431	1.80
Er	12.4	8.70	82	Ni	1.08	1.02	15.7
Gd	10.6	6.56	73.6	Mo	1.05	7.38	135
Ho	9.15	10.9	96.5	La	0.915	2.09	65
Hf	88.60	5.43	40.2	Ce	0.823	2.59	73.7
Y	6.15	30.2	262	Zr	0.756	7.38	132.5
V	5.77	38.7	1010	W	0.710	2.12	35.5
Dy	5.71	10.6	106.2	Ta	0.596	3.32	26.5
Tm	4.51	27.8	101	Pt	0.382	0.051	0.31
Mn	4.31	2.79	59.5	Pb	0.230	3.45	4.53
Yb	4.27	11.4	123.5	Cu	0.227	0.656	4.97
Ti	3.66	16.7	405	Ag	0.181	0.319	0.807
Be	3.37	33.3	812	Pd	0.148	0.105	7.4
In	3.34	63.7	217	Zn	0.145	0.354	2.76
Co	3.20	0.953	25	Sn	0.070	0.572	2.13
Pr	2.46	9.85	103	Au	0.045	0.059	0.24
Nb	2.16	12.3	148	Cd	0.040	0.055	0.408
Re	2.14	8.21	108				

emission can be obtained by tracing the dependence of K^+ on the atomic number Z of the target element. Fig. 1.8 shows in semilogarithmic scale the dependence of K^+ on atomic number Z. Here log $K^+_{i\,rel} =$ log (I^+/I^+_{Fe}) and Fe has been taken as an internal reference sample. Analysis of these data allows to reveal a certain periodical dependence of K^+_i on atomic number of the target, the character of this dependence varying slightly with the primary ion species.

This phenomenological information may be useful for understanding effects of various parameters characterizing the properties of bulk target and isolated atoms sputtered from this target. The following parameters may be considered here from general point of view: electron work functions of the surface being bombarded; density of electron states at the Fermi level; $N(E_F)$; interatomic binding energy H; free atom ionization potential V_1; mean energy of sputtered particles E.

Plotting of K^+ vs various parameters indicates the existence of certain correlations, e.g. K^+ variation with Z corresponds well to variation of reciprocal ionization energy $1/V_i$. However, the effect of this parameter alone, although very important, cannot explain all the peculiarities in the dependence $K^+_i = f(Z)$. For example, in the group of elements Sc, Y, and La, K^+ is decreasing while V_i and . remain practically the same [63]. The importance of $N(E_F)$ can be valued if one compares the dependence $N(E_F)$ for d-transition metals on the total number of outer $d+s$ electrons with $K^+_i(Z)$ periodicity. Data for $N(E_F)$ are taken from [64] and are shown in Fig. 1.9. It can be seen from Figs 1.8 and 1.9 that variation of K_i^+ in the periods for Ar^+ ion bombardment is in good qualitative agreement with variation of $N(E_F)$. Decrease of $N(E_F)$ in the groups with growing main quantum number corresponds to the drop in the emission intensity.

All theories of sputtering consider the interatomic binding energy as an important parameter [65]. Therefore it is useful to compare the dependence of sublimation energy H on the atomic number Z and $K_i^+(Z)$. As in previous cases there is no unambiguous correspondence between changes in the SIE intensity and the binding energy. The binding energy in a series of elements is at first increased but then becomes lower, while the K_i^+ has a general tendency to become smaller as Z increases. However for the elements of one group a distinct dependence between these characteristics may be observed. For the subgroups I B, II B, and III B the K_i^+ is an increasing function of the binding energy, but for the subgroups IV B and III A – VII A it is a decreasing function.

Analysis of $K_i^+ = f(Z)$ shows that the number of secondary ions cannot be connected with the action of any single parameter. A number of factors must be taken into account, their influence being not constant but depending on the nature of target and primary ions.

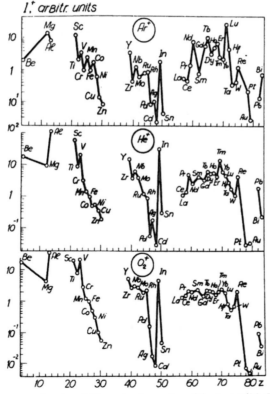

Figure 1.8 Relative secondary ion yields vs atomic number of the target (8 keV Ar^+, He^+, and O_2^+ primary ion bombardment).

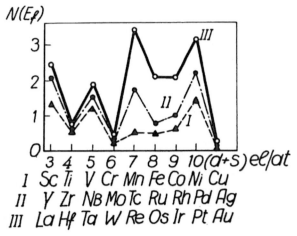

Figure 1.9 The density of electron states at the Fermi level for metals of first, second, and third large periods vs number of outer $d + s$ electrons in the atom.

The above correlations being taken into account, one can write in a general form

$$K_i^+ = f[V_i, H, N(E_F), (V_i - \varphi)] \tag{1.8}$$

and to find an exact empirical function that would be satisfactory for the approximation of the dependence $K_i^+ = f(z)$. For argon ions we have

$$K_{i\,rel}^+ = A \frac{N(E_F)}{V_i - \varphi >} \exp\left[-B(V_i + 0.1\,H)\right] \tag{1.9}$$

where $A = 1.73 \times 10^{-7}$ eV and $B = 2$ eV^{-1}.

From the comparison of data on secondary ion energy distribution with those on the SIE coefficient variation depending on the target atomic number, it may be assumed that the probability of ion ejection is directly connected with the average energy of sputtered particles. Hence this parameter must be included in the empirical formula (1.9). This gives the following empirical formula for the relative SIE coefficient:

$$K_{i\,rel}^+ = 7.7 \times 10^{-5} \frac{\bar{E}N(E_F)}{V_i - \varphi} \exp\left[-2(V_i + 0.1\,H)\right] \tag{1.10}$$

Now we are in a position to explain some features in the SIE of particular elements. Elements in Group III that have one $(n-1)$ d-electron (subgroup III A) or one np-electron (subgroup III B) are characterized by low values of V_i and H, but have a high density of states $N(E_F)$. Therefore these elements are characterized by a high value of K_i^+. The increased SIE intensity of elements in the subgroup V A (V and Nb) as compared to that of elements in the subgroup IV A (Ti and Zr) is also the result of higher $N(E_F)$ values and lower ionization energy.

With Ar$^+$ bombardment an anomalous low emission was found for Pd. The reason may be a rather high ionization potential for Pd atoms, due perhaps to the filling of the $4d$-shell owing to the absence of $5s$-electrons. A sharp increase in Re emission comparing to W is also due to the decrease in the binding energy between the outer electron and the atom [or due to still larger lowering of the parameter $(V_i - \varphi)$ and the increase in the density of states $N(E_F)$].

Anomalously high ion emission has also been established for Hf in the IV A subgroup. This is connected with the anomalously low value of the ionization energy and the sublimation energy H, which apparently plays a dominant role in the ionization process.

The highest secondary ion emission intensity for Ar^+ ions gives Lu. This element has a completely filled $4f$-shell with one $5d$-electron and a high density of electron states at the Fermi level. It also has the lowest ionization potential. It should be mentioned that for this element the K_i^+ value from the surface oxides is smaller than from pure metal. Hence the Lu^+ high emission cannot be caused by the oxidation effect due to oxygen in residual gases.

Investigation of 13 lanthanides with the aim to define the effect of occupation of the inner $4f$-shell has not revealed any definite dependence on atomic number. It can only be said that with Ar^+ bombardment the $K_i^+(Z)$ variation agrees qualitatively with variation of inter-atomic binding energy for this series. From comparison of SIE intensity variation for lanthanides as a function of Z for different primary ions, the following facts are found: when changing from Ar^+ to He^+ or O_2^+ the 'pulsation' amplitude of K_i^+ grows while the level of the average intensity becomes lower.

It may be seen from Table 1.3 that for most elements the transition to He^+ ions gives higher ion currents than in the case of Ar^+. Since the sputtering coefficient for He^+ is much smaller, it may be concluded that particles sputtered by lighter ions have a higher ionization efficiency. This is an important observation since the use of light ions, with which sputtering yields are low and the surface erosion is minimum, may enlarge the applications of SIMS for analytical purposes. Hydrogen ions are the lightest and most promising from this point of view, but so far little attention has been paid to this possibility, probably because of the danger of expected hydride interferences in mass spectra. The first systematic comparative study of various aspects of SIE from transition metals bombarded with hydrogen and argon ions [66, 67] has shown very high ionization efficiency for hydrogen ion bombardment. The measured differential yields of SIE from 17 transition metals of the first, second and third long periods are presented in Fig. 1.10(a). It is seen that with H^+ and Ar^+ primary ions the secondary ion yields vary with the atomic number Z of the metal in a similar manner, and K_H^+ and K_{Ar}^+ measured for the same element yield similar figures. Moreover, for a number of metals (e.g., Fe, Cu, Zn, Ag, and Au) the SIE rate is much higher when induced by hydrogen as opposed to argon ions [Fig. 1.10(d)]. This is an interesting finding in view of the fact that the sputtering yields of metals bombarded by argon ions, S_{Ar}, are higher than yields with hydrogen, S_H [Fig. 1.10(b, d)]. This effect can be accounted for by the extremely high ionization

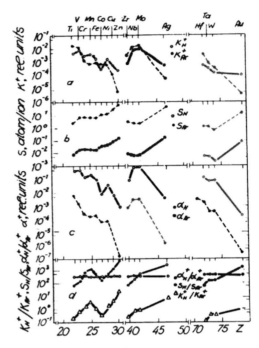

Figure 1.10 Secondary ion emission yields, K^+ (a); sputtering yield, S (b); ionization probability, α^+ (c); ratio of these values (d) — for pure metals bombarded with Ar and H ions.

efficiency of sputtered atoms produced by hydrogen ion bombardment. For the metals in question the ionization efficiency R^+_H exceeds R^+_{Ar} by a factor of $10^2 - 10^4$. The experimentally detected excess of R^+_H over R^+_{Ar} suggests the profound effect of a change in the physico-chemical properties of the surface as a result of its interaction with hydrogen ions [66]. These and similar effects are very important in practical SIMS and are considered in more detail in the next section.

1.7. Effect of sample oxidation on SIE

As soon as a high sensitivity of SIE to the surface state and its dependence on vacuum conditions were discovered, many laboratories began systematic studies of the effect of various gases. The results are summarized in [2]. Chemisorption of oxygen and the oxidation process produce the strongest effect on secondary ion yield. Investigation of the oxygen pressure effect on SIE regularities gave rise to a new scientific direction which deals with the use of this phenomenon for the study of gas interaction with solid surfaces. The model of so-called chemical emission was proposed to explain a significant growth of K^+_i as a result of O_2 action [57, 68, 69]. It should be noted that the oxidation effect has its pronounced manifestation in kinetic dependences of the secondary ion current, that is in variations in the magnitude of the secondary ion current during bombardment of the target surface by inert or active gas ions. Several authors have shown that the secondary ion emission intensity varies strongly during bombardment, a stable level being reached only after a definite time of surface sputtering. In [70] this was demonstrated using the example of an Al target, while in [71] the example was Fe.

Systematic investigation of the effect of the physico-chemical state on SIE kinetics was performed in our laboratory. A large number of objects were studied using Ar^+ and O_2^+ bombardment. The physico-chemical state of metal surface under study was changed mainly through controlled oxidation and adsorption processes at room temperature and through the dose of ion irradiation. Oxide films as thick as several nm were produced in the ambient atmosphere using the known method [72]. The effect of pure oxygen was studied by admission of gas into the vacuum chamber under controlled pressure. The surface composition and its physico-chemical state were controlled by AES and the appearance potential spectroscopy methods.

As mentioned above, secondary ion current depends strongly on the degree of surface coverage by oxygen molecules, or on surface oxidation. This may be most clearly observed in kinetic curves $K_i^+(\tau)$ recorded during an oxidized surface bombardment.

Investigation of SIE kinetics makes it possible to reliably establish the conditions under which stable values of secondary ion current can be obtained. This information is valuable from the point of view of a possible SIE application for the study of oxidation and adsorption processes using comparatively high ion current densities. The SIE kinetics have been studied as a function of target material, density of Ar^+ ion current J_0^+ and the exposure dose D of the sample in an oxygen atmosphere under different pressures P.

The presence of an oxide layer on a sample original surface is responsible for a sharp increase in the intensity of ion current (the oxidation peak), which is usually observed at the initial moment of ion bombardment. With further sputtering of the oxide layer the emission intensity is decreased. After a definite time τ_{st} the intensity of the secondary ion current becomes stable, indicating the achievement of a dynamical equilibrium between the rate of adsorption of residual gas at the surface and the rate of its removal under the action of ion bombardment. If, after τ_{st} has been reached, the ion beam is interrupted for some time, the kinetic dependence $K_i^+(i)$ will be defined by the action of the residual atmosphere oxygen on the surface (an adsorption peak is observed).

The main types of kinetic curves obtained at the bombardment of different metals by Ar^+, He^+, and O_2^+ ions with a constant current density $J_D^+ = 1.5$ mA cm^{-2} are shown in Fig. 1.11. Most common are the curves of type a. The shape of the curve is typical for initial and adsorption peaks when elements with a high oxygen affinity (Sc, Ti, V, Cr, Nb, Mo, Ta and some lanthanides) are bombarded by Ar^+. It is also characteristic of the adsorption peak when He^+ ions are bombarding the targets of Ta, Mg, Al, Mn, and some lanthanides. With O_2^+ bombardment, type A curves are not as common (Mg, Fe, Co, Ta). The kinetic curves of the type under consideration are characterized by a fast run of the transition process, i.e. by small values of τ_{st} (9 – 40 s).

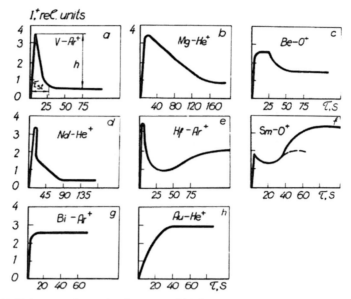

Figure 1.11 Main types of secondary ion current kinetic curves.

The kinetic dependence of type b is obtained when Al, Y, In, Ti and lanthanides are bombarded by Ar^+ ions, and when Be, Mg, Sc, V are bombarded by He^+ ions. In this case the τ_{st} value amounts to $45 - 115$ s due to the low sputtering rate of stable oxides. For O_2^+ the type b curves have a long period of the ion current increase and decrease.

The kinetic type d curve is a superposition of the a and b curves and is usually observed with He^+ bombardment of elements with high oxygen affinity. A sharp spike here (of type a) is due to the emission from a thin layer which has been formed during the exposure, while the next intensity drop (of type b) results from the presence of the oxide which had not been completely sputtered.

In case when the ionization probability for the oxide sputtering is lower than that for the sputtering of a pure element, dependences of d and f types are observed. Type e kinetics may be observed at the initial moment of Ar^+ bombardment of Hf and W targets, after a prolonged He^+ bombardment of a Ru target, and with O_2^+ bombardment of Mn. The surface peak of type f refers to Ar^+ bombardment of Lu and to the system $O_2^+ - Sc$, La, Cd, Nb, Sm.

The adsorption peak of type f is also observed for the system $Ar^+ - Ag$ and $He^+ - Cu$, Zn, Pd, Cd. There is no surface peak for elements with a high oxygen affinity under O_2^+ bombardment. If the metal surface is covered by a compound that is hard to sputter, the emission from which is lower than that from the target itself, type h kinetics ($Ar^+ - Zn$, Zr, Hf, Pt; $He^+ - Ag$, Hf, W, Au; $O_2^+ - Zn$, Au) are observed.

Analysis of a large number of SIE kinetic dependences have made it possible to establish a periodic dependence of the peak relative height [$h = (I_{peak} - I_{st})/I_{st}$] on the element atomic number, with Ar^+ bombardment. The surface peak value drops to zero as the atomic number rises, within the limits of large periods. Within groups the decrease of this parameter is also observed.

From the point of view of reliability of analysis for ions of bulk origin, it is of interest to consider the nature of adsorption peaks appearing after repeated bombardment of the target surface. Such peaks result from residual oxygen adsorption on the target surface, which is cleaned in the course of preliminary bombardment.

Under otherwise identical conditions, the intensity of this peak depends on the properties of target material and the oxygen partial pressure in the vacuum chamber. The binding energy E of a chemisorbed gas with electron affinity S may be found from [73]:

$$E = (S - \varphi)e + e^2/4d \qquad (1.11)$$

The increase in the electron work function within the series of elements leads to a lower binding energy and hence to a lower adsorption energy. Therefore the largest adsorption peak is observed for elements with a high oxygen affinity and a small work function. However in the case under consideration we omitted the effect of preliminary bombardment of the surface by inert gas ions on the adsorption process. It has been found [74] that implantation of inert gas ions greatly affects the oxidation behaviour of metals in ambient atmosphere. This factor was later taken into account in the study of the effect produced by doses D_0 of Ar^+ and He^+ irradiation on the electron work function and on the magnitude of the surface and adsorption peaks [75–77]. It has been established that variation of work function with irradiation dose characterizes the surfaces cleaning process, whose completion coincides with the appearance of stable K_i^+. Further increase of the irradiation dose causes no changes in $e\varphi$, indicating that the element is pure [77].

All the above data refer to the pressure range around 10^{-6} Torr most widely used in practical SIMS. More accurate measurements have demonstrated that chemical enhancement can be observed even in the 10^{-8} Torr range. The most 'clean' data, referring to clean surfaces in the 10^{-9} Torr pressure range, were obtained in [20]. One of the most important results of this study is that relative secondary ion yields of clean elements differ by only two orders of magnitude. Therefore one must be very careful when trying to check the validity of this or that theory by comparison with existing experimental results, and rely only upon data corresponding to conditions close to ideal ones used in physical models.

1.8. Polyatomic ion emission

A mass spectrum of secondary ion emission contains, as a rule, ions of the following origin: target atoms; bulk contaminations of the target substance; ions of the same nature as primary beam ions; molecules and molecular fragments of chemical compounds present in the target surface; atoms, molecules, or molecular fragments absorbed on the target surface. For measurements in high or ultra-high vacuum with an atomically clean surface the mass spectrum is composed of singly or multiply charged ions and polyatomic ions from target atoms (cluster ions). The study of cluster ion emission gives valuable information on various surface phenomena at the solid – gas interface: adsorbtion, oxidation, catalysis, sputtering, ionization, molecular stability, etc. [2, 3, 79–81]. A connection between the emission of such ions and the physico-chemical state, structure, and phase composition of the target has been established [2, 82, 83].

The charge state and energy of primary ions in the several keV range produce little effect on the emission of cluster ions, but an increase in the mass of the bombarding particles gives rise to considerable enhancement of this process [84, 85]. As an example, the relative intensity of Al_n^+ ions sputtered from an Al target by Xe^+, Ar^+, and He^+ ions is shown in Fig. 1.12. It is seen from this figure that transition from Xe^+ to He^+ strongly decreases the emission of Al polyatomic ions, and Al_n^+ ions with $n > 4$ are not observed in the spectra for He^+ bombardment. The dependence of cluster ion emission on the nature and pressure of residual gases has been studied in detail in [2, 79, 81, 85]. It has been shown that intensity of emission of matrix and surface cluster ions may be controlled by variation of the partial pressure of electronegative gases in the sample chamber.

The energy distribution of cluster ions is, as a rule, narrower and peaks at lower energies than the energy distributions of monoatomic ions (52, 87–90] (Fig. 1.13). Formation of cluster ions from a pure metal surface depends on the electron structure of the target and the isolated target atoms. It has been established that for transition metals of the first large period the emission intensity of cluster ions monotonously decreases with the growth of atomic number of ions of the Me_n^+ type. It has been found [86, 87, 90, 91] that for such metals as Zr, Mo, Ta, W, Nb, and Cu the dimer – monomer ratio, $K^+_{Me_2}/K^+_{Me}$, exceeds unity. For instance, the Nb_2^+ yield is twice than for monomer ions. The following explanation was proposed [96] for the observed relationship. When two excited atoms leave the surface as a molecule their excitation

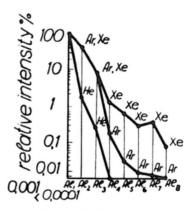

Figure 1.12 Relative intensity of Al_n^+ ions emitted from Al target bombarded by He^+, Ar^+, and Xe^+ ions.

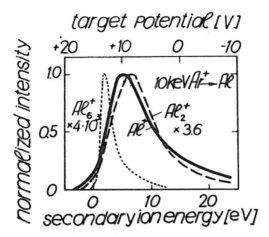

Figure 1.13 Energy spectra of Al_n^+ ions without correction for variation in emittance of the secondary ion source [20].

energies are added. Since the ionization energy of the molecule differs little from that of an isolated atom, the ionization probability may be larger in the first case than in the second. This effect is more pronounced the higher is the ionization probability of atoms, but should not be manifested, for example, for Au or Cu, where atoms possess high ionization potentials. However it is shown in [87] that $K_{Cu_{2n}}^+ / K_{Cu_{2n+1}}^+$ varies with n, and this dependence is of a periodic character depending on n. To explain the mechanism of this phenomenon, a model has been developed which takes into account the effect of the electron structure of the target material and the influence of the emitted cluster structure on the probability of its ionization. The essence of this model is as follows. The dissociation energy of the cluster may be expressed as the difference between the electron exchange energy obtained when atoms are approaching from infinity to the distance of the cluster size, and the Coulomb energy of nuclear repulsion. The number of bonds Z_n for atom pairs with a minimum separation d may be calculated using Fig. 1.14; where possible configurations $N(nZ_n)$ are shown for the same n and Z_n. The following expressions have been deduced to determine the Coulomb interaction energy ϵ_c^+ for Cu_n^+ ion formation:

$$\epsilon_c^+ = - E_j + (n - I)\omega^+ + K + \beta^+ \qquad (1.12)$$

Here E_j is the energy necessary to remove an outer s-electron; ω^+ and β^+ are the hopping integrals; and K is a parameter accounting for the complex configuration.

In the calculation of molecular level energies, the approximation proposed in [92] was used. For calculation of the Coulomb repulsion energy, ϵ_c it is assumed to be negligibly small for atoms separated by distances larger than d, and proportional to Z_n neighbouring atoms. According to Equation (1.12), with the assumption that ω^+ and β^+ do not depend on n and the cluster configuration, the maximum ϵ_c value corresponds to maximum K. Therefore

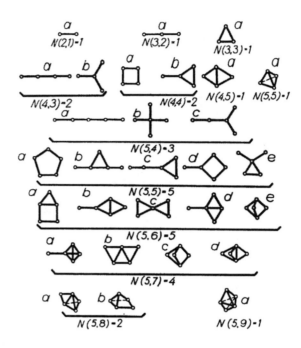

Figure 1.14 Different cluster configurations at the same values of n and Z_n. a – e are the configuration types.

from all the proposed configurations (Fig. 1.12) the most stable are those for which K is the largest. In particular, for the same n, the polyatomic ion with a large number of bonds Z_n may turn out to be less suitable if its configuration does not correspond to the maximum K. For interaction energy calculation only strong bonds of outer d-electrons may be considered. The effect of d-electrons on the stability of a complex at large n is assumed to be insignificant. Thus the proposed model is not intended for accurate determination of geometrical characteristics of stable ion forms for each n value, but may be useful for evaluation of the stability of a particular form or, in other words, for evaluation of the 'survival' probability of emitted ions.

Since the ionization probability of sputtered complex depends on the ionization energy Δ_n^+, it has become possible to calculate this parameter, with the aim of explaining the periodic dependence of $K_{i(n)}^+$ [93]. The ionization energy was found as the difference between the energy of the last energy level filled, and zero energy. It turns out that the dependence $K_{i(n)}^+$ varies in the same sequence as the dependence of $(-\Delta_n^+)$ on n. According to ideas developed in [94–95], this implies that the ionization probability and stability of complexes, that are characterized by values of n, are higher than those for ions with even indexes. This theory can also explain the monotonous character of the emission intensity decrease for polyatomic ions at larger n, this decrease being experimentally established for transition metals of the first large period [80] and for light elements (Li and Be) [96].

It should be emphasized that the emission of polyatomic and molecular ions is the least studied field and makes an important separate problem, solution of which opens unique possibilities for use of SIE in the study of solids.

1.9. Mechanism of ionization of the ejected atoms

The wide variety of experimental data on secondary ion emission needs to be explained theoretically if physical predictions are to be obtained and quantitation procedures developed. There are a number of possible approaches to explaining the ionization process, although none have yet proven very satisfactory [1–3, 97, 99]. These approaches, or models, can be classified into three groups, differing essentially from each other.

Kinetic models. In these, the ionization mechanism is essentially determined by the kinetic energy of the ion being sputtered and associated with inelastic energy losses during the collision cascade and with local damage of the surface during the impact of the primary ion.

Surface effect models. The secondary ions are considered to be formed by electron exchange between the atom being sputtered and the metal surface or by bond-breaking effects in ionic crystals and other dielectrics.

Thermodynamic models. Some kind of local thermal equilibrium at the surface being bombarded is assumed.

The kinetic model describes secondary ion emission as a three-step process:

(a) penetration of a primary ion into the target and emission of the secondary ion as a result of collision cascade;

(b) creation of inner-shell electronic excitation ('deep holes') during collisions between the target particles;

(c) emission of target particles as excited neutral species (for sufficiently long electronic relaxation times). De-excitation outside the metal by an Auger process yields a positive secondary ion and Auger electron(s) [100].

This model is in qualitative agreement with experimental data for some metals, but it is far from being general for various types of target.

Close to this is the auto-ionization model [101, 102], in which creation of an inner level electronic excitation is followed by transition of two or three conduction-band electrons into the energetically equivalent auto-ionizing states of an atom at or near the target surface, with subsequent relaxation of the auto-ionizing states in vacuum by an Auger effect, yielding the secondary ion and an Auger electron(s). The sputter ion yield S^+ is expressed as $S^+ = P_h P_A S^0$, where P_h is the probability of the sputtered particle making a deep hole near the surface, P_A is the probability of an electron making the transition into an auto-ionizing state, and S^0 is the sputtering yield. For comparison with an experiment, P_h and S^0 are assumed to be independent of the element, so that $S^+ = P_A$, but for the calculation of P_A detailed knowledge of the electronic structure of the metal and of the free atom is required.

With some simplifying assumptions, variation of ion-yield from element to element for many transition metals can be represented correctly within a factor of better than two [101]; a number of matrix effects can be also explained [102].

In the surface effect models all the processes relevant to ionization of a sputtered particle are assumed to take place directly at or immediately in front of the sample surface. A particle, during ejection from the surface, must change its electronic structure from that of the bulk substance to that of the free atom. This may lead to electron transitions from the sputtered atom back to the target, with ion formation as a result. The probability R^+ of getting a sputtered particle in an ionized state at a large distance from the surface is then equal to ratio of emitted ions to total emitted particles, $R^+ = S^+/(S^0 + S^+) \approx S^+/S^0$. Sroubek *et al.* [103] assume the flight-time of the particle through the surface region to be short, and calculates R^+ according to a 'sudden' approximation approach:

$$R^+ \cong S^+/S^0 = [(I - \Phi) \pi \cdot V/n \Delta^2 a]^{1/2} \qquad (1.13)$$

where a is the width of the surface zone, I the ionization energy of the particle, V the velocity of the particle, Φ the work function of the metal, n an integer, and Δ the width of the broadened 'adatom' ground state immediately at the surface. Since Δ is not calculated explicitly it may vary strongly from element to element, accounting for the large variations in sputter ion yield. Detailed comparisons with experimental data are not available, so this model has a qualitative character.

Within the adiabatic approximation Schröeer *et al.* [104] calculated the same transition probability and obtained

$$R^+ = S^+/S^0 = \frac{B^2}{(I - \Phi)^2} [\hbar \bar{v}/a \, (I - \Phi)]^n \qquad (1.14)$$

where $\bar{v}$ is the mean velocity of sputtered particle, B is the surface binding energy, and a and the exponent n are considered as free parameters and are fitted to experimental sputter ion yields. It should be noted that numerical values of a and n depend strongly on the experimental conditions, and ion-yields of metals can generally be predicted to within a factor of two to four. Gries and Rüdenauer [105, 106] have modified the Schröer theory by fitting n to experimentally observed secondary ion energy distributions and calculating S^0 and V analytically from Sigmund's sputtering theory [107]. Cini [108] has given a different derivation of R^+ and found that owing to different electronic structure of an adatom near a metal surface, no unique velocity dependence of R^+ may exist for all elements.

Antal [109] considers a target particle moving through the metal lattice as an interstitial representing a local excess of positive charge, exactly balanced by an 'electron cloud' moving with the particle, thus forming a quasi-atom. A de Broglie electron wave attached to this electron cloud is partially reflected when the particle crosses the potential wall formed by the sample surface, and stripping of electrons occurs. The ionization probability R^+ of sputtered atom can therefore be equated to the quantum mechanical reflection coefficient of the electron wave at the surface potential barrier:

$$R^+ = [(I - \beta)/(I + \beta)]^2, \tag{1.15}$$

$$\beta \cong [(2I - \Phi)/(E_F + V_a)]^{1/2} \tag{1.16}$$

where E_F is the Fermi energy and V the screening potential of the moving electron cloud.

Without use of fitting parameters, calculated ion yields for many pure metals agree with experimental data within a factor of two, with the exception of the noble metals for which yields are over-estimated by some orders of magnitude.

The well-known local thermal equilibrium (LTE) model by Andersen and Hinthorne [110] is the most widely used one in actual analytical work, and the model most widely criticized for lack of physical argumentation. It is based on the assumption that a 'surface plasma' exists on an ion-bombarded sample in which neutrals, ions, and electrons are in a state of local thermodynamic equilibrium which can be described by the Saha–Eggert equation:

$$\frac{\eta^+ \cdot \eta^e}{\eta^0} = \left(\frac{2\pi \, mkT}{h^2} \right)^{3/2} \frac{2Z^+}{Z^0} e^{-(I-\Delta E)/kT} \tag{1.17}$$

Here, η^+, η^0, η^e are the volume concentrations of ions, neutrals and electrons, respectively; Z^+ and Z^0 are the temperature dependent partition functions of ions and neutrals, respectively; h and k are Planck's and Boltzmann's constant, respectively; m is the mass of the electron, I the first ionization potential of the particle, and ΔE the 'depression' of the ionization energy due to collective plasma effects; and T is the 'temperature' of the surface plasma.

The ratio of detected secondary ion currents from elements A and B in a multi-element sample is assumed to be equal to the volume density of the respective charged species in the surface plasma: $i_A^+/i_B^+ = \eta_A^+/\eta_B^+$. Therefore concentration ratios for all elements in a multi-element sample may be calculated from experimentally measured secondary ion currents using the Saha–Eggert equation, provided the plasma parameters T and η^e are known. The state of the plasma, characterized by T and η^e, is determined from the ion emission of at least two internal standard elements in the sample by means of a mathematical algorithm described in the literature (see Chapter 3). Nevertheless, while accepting the remarkable practical success of this method in predicting absolute sample composition, the many conceptual difficulties of the method must be borne in mind. It turns out that in the Saha–Eggert equation, a number of parameters may be changed rather arbitrarily, and rigorous simplifications may be made, without markedly affecting the quality of analysis. It is probable that Werner's interpretation of the ion emission process [111] as a series of uncorrelated statistical emission events leading to a Saha–Eggert type of equation, may be a more adequate description of the phenomenon.

In any of the models considered above, variation of the electronic structure of the ejecting atom and the solid are not considered in any detail. Another problem which must be taken into account is that the electronic structure of the solid must surely change as the collision cascade evolves in time. For example, after the impact of the primary ion, initial periodicity in a crystal is distorted completely and the atoms of the target more closely resemble a liquid. The atoms in the collision cascade are moving with energies of several eV per atom; hence atom–atom distances can vary over a much wider range than are normally found in amorphous materials. This disorder may have the effect of localizing the itinerant electrons of a metal or of creating new states with energies within the bandgap of semiconductors or insulators. What is therefore really required is a formalism where electronic motion and nuclear motion are determined in time together.

This approach offers considerable promise in understanding the important factors influencing surface ionization [112]. Both electronic and nuclear motion are considered using a Hamiltonian, which is separated into a part that describes the motion of the atomic nuclei and a part that describes the motion of the electrons. The nuclear positions are determined using a molecular dynamics calculation on a large ensemble of atoms to compute actual nuclear positions as they change in time subsequent to the primary ion impact [113].

In this way the appropriate time-dependent coordinates for solution of the electronic part of the problem are provided [112].

The ionization probability R^+ for an adatom ejected from a surface may be determined by solving the time-dependent Schrödinger equation for the system, using the time-dependent electronic Hamiltonian

$$H = H_{at} + V[r(t)] \tag{1.18}$$

Here, H_{at} is an atomic Hamiltonian, and V is time or position-dependent interatomic electronic interaction integral. The coupling interaction neglects electron–electron interaction, so no two electron or Auger type processes are considered. The solutions to H_{at} are given by

$$H_{at}\,\Phi_k = \epsilon_k\,\Phi_k \tag{1.19}$$

where the Φ_k's are the orthonormal, atomic wave functions for the atoms in the solid.

To solve the Schrödinger equation using the Hamiltonian in Equation (1.18), one then seeks time-dependent molecular wave functions ψ_i where

$$H\,\psi_i = i\hbar\,\frac{\partial\,\psi_i}{\partial t} \tag{1.20}$$

Using the tight binding approximation, the electronic levels in the solid may be expressed as a linear combination of atomic orbitals Φ_k:

$$\psi_i = \sum_k \Phi_k C_{ki}(t)\exp(-iE_i t/\hbar) \tag{1.21}$$

where an arbitrary phase factor has been included. The expansion coefficients are time-dependent, since the position of the nuclei will be changing during the evolution of the collision cascade. Combining Equations (1.20) and (1.21) and integrating over the electronic coordinates yields the following set of coupled equations;

$$i\hbar[dC_{ki}(t)/dt] = \sum_{j \neq k} <\Phi_k|V|\Phi_j>C_{ji}(t) + (\epsilon_k - E_i)C_{ki}(t) \tag{1.22}$$

These equations of motion for the $C_{ki}(t)$'s can be integrated simultaneously with the classical equations for the nuclear motion. As it was in the two-level case, the coupling matrix element or hopping integral is given by

$$<\Phi_k|V|\Phi_i> = V[r(t)] = V_{ki}\exp[-\lambda_{kj}(r_{ki} - r_{ki}^e)]; \quad r_{ki} \neq 0,$$
$$k \neq j. \tag{1.23}$$
$$r_{ki} = 0,$$
$$k \neq j. \tag{1.24}$$

The value of R^+ may be calculated from the coupling coefficients by choosing some atomic energies ϵ_k and parameters for $V[r(t)]$ and then solving for the molecular energies E_i at time $t = 0$ by setting Equation (1.22) equal to zero. Then the nuclear motion of the total system must be determined by the molecular dynamics method [112]. At each time step the new nuclear positions serve as input to the coupling matrix element. The coupled equations of motion for the $C_{ki}(t)$'s are simultaneously integrated with the nuclear positions and velocities.

R^+ is evaluated at the end of the collision process by projecting the adatom state Φ_a on all states that were originally unoccupied. R^+ is thus given by

$$R^+ = \sum_i^{unocc} |C_{ai}(t = \infty)|^2 \tag{1.25}$$

The numerical analysis for this expression is beyond the capacity of most present-day mainframes. Currently, the solid is approximated using a liner chain of atoms [114] and by a cluster of five atoms with the adsorbate atom placed on its surface [115]. Another approach involves monitoring the nuclear motion of a large collection of particles (~ 300 or more) while only selecting a few (5–10) to follow electronically.

The incorporation of nuclear motion into ionization theory is the way to obtaining physically reasonable results. Sroubek found [114, 115] that by varying the ionization energy of the ejecting adsorbate atom, R^+ values could be produced ranging from 10^{-1} to 10^{-5}, which are much more reasonable than those obtained from lattices that are not allowed to move. Although the development of this procedure is in its very early stages, it appears that all of the fundamental ingredients are available to construct a general understanding of the ionization phenomenon [112].

References

1. Veksler, V. I. (1970) *Secondary emission of atomic particles*. FAN, Tashkent (in Russian).
2. Cherepin, V. T. and Vasil'ev M. A. (1975) *Secondary ion–ion emission from metals and alloys*. Naukova dumka, Kiev, (in Russian).
3. Veksler, V. I. (1978) *Secondary ion emission of metals,* Nauka, Moscow (in Russian).
4. Cherepin, V. T. (1979) Secondary ion mass spectrometry. *Elektron. prom-st'* 1(2) 17–34, (in Russian).
5. Mc Hugh, I. A. (1979) Secondary ion mass spectrometry In: *Metody analiza poverkhnostey.* pp. 276–341, Mir, Moscow (in Russian).
6. Veksler, V. I. (1980) The ionic component in the sputtering of metals. *Radiat. Eff.* 51, 129–170.
7. Cherepin, V. T. (1981) *The ion probe.* Naukova dumka, Kiev, (in Russian).
8. Cherepin, V. T. and Vasil'ev, M. A. (1982) *Methods and instruments for analysis of surface of materials.* Naukova dumka, Kiev (in Russian).
9. Turner, N. H. and Colton, R. J. (1982) Surface analysis: X-ray photoelectron spectrometry Auger electron spectroscopy and secondary ion mass spectroscopy. *Anal. Chem.* 54, 293R–322R.
10. Turner, N. H., Dunlap, B. I., and Colton, R. J. (1984) Surface analysis: X-ray photoelectron spectroscopy, Auger electron spectroscopy and secondary ion mass spectrometry. *Anal. Chem.* 56, 373R–416R.
11. Stanton, H. E. (1960) On the yield and energy distribution of secondary positive ions from metal surfaces. *J. Appl. Phys.* 31, 678–683.
12. Hennequin, J. F. (1968) Distribution energetique et angulaire de l'emission ionique secondaire. III. Distribution angulaire et rendements ioniques. *J. Phys.* 29, 957–68.
13. Jurela, Z. and Perovic, B. (1968) Mass and energy analysis of positive ions emitted from metallic targets bombarded by heavy ions in the keV energy region. *Can. J. Phys.* 46, 773–778.
14. Bukhanov, V. M., Yurasova, V. E., Sysoev, A. A. *et al.* (1970) Ion component in the cathode sputtering of single crystal copper. *FTT* 12, 394–397 (in Russian).
15. Yurasova, V. E., Sysoev, A. A., Samsonov, G. A. *et al.* (1973) Spatial and energy distribution of secondary ions produced by ion bombardment of single crystals. *Radiat. Eff.* 20, 89–93.
16. McCoughan D. V., Sloane R. H., and Geddes J. (1973) An apparatus for study of secondary ions from ion bombardment of a metal surface. *Rev. Sci. Instrum.* 44, 605–610.
17. Yurasova, V. E. (1975) Emission of atomic particles at ion bombardment of single crystals. Thesis of doctoral science dissertation, MGU, Moscow (in Russian).
18. Bayly, A. R. and MacDonald, R. I. (1977) The energy spectra of secondary ions emitted during ion bombardment. *Radiat. Eff.* 34, 169–181.
19. Pleshivtsev, N. V. (1968) *Cathode sputtering.* Atomizdat, Moscow (in Russian).
20. Wittmaack, K. (1975) Energy dependence of the secondary ion yield of metals and semiconductors. *Surf. Sci.* 53, 626–635.
21. Fogel, Ya.M. (1976) To the question of the choice of the current density magnitude for the primary ion beam when processes on solid surface are studied by SIIE method. *Zh. Tekh. Fiz.* 46, L 787–L788 (in Russian).
22. Wittmaack, K. (1976) Current density effects in secondary ion emission studies. *Nucl. Instrum. Meth.* 132, 381–385.
23. Vasil'ev, M. A., Goncharenko, A. B., Chenakin, S. P., and Cherepin, V. T. (1979). The mechanism of effect of the inert gas primary ion nature on the ionization probability of sputtered atoms. *Metallofizika* 1, 101–105 (in Russian).
24. Fogel, Ya. M. (1967). Secondary ion emission *UFN,* 91, 75–112 (in Russian).
25. Koval', A. G. (1977) SIMS use for the study of processes on solid surface and in the bulk In: *Diagnostika poverkhnosty ionnymi puchkami,* p. 27, Uzhgorod: Uzhgorodsky Un-t, (in Russian).
26. Vasil'ev, M. A., Ivaschenko, Yu. N., and Cherepin, V. T. (1973). The effect of a metal target temperature on the secondary ion–ion emission *Metallofizika,* 53, 64–67 (in Russian).
27. Yurasova V. E. (1983) Emission of secondary particles during ion bombardment of metals in the phase transition region—Part 1. Sputtering. *Vacuum,* 33, 565–578.
28. Staudenmaier, G. (1973) Angular dependence of cluster sputtered from a tungsten single crystal surface. *Radiat. Eff.* 18, 181.

29. Vasil'ev, M. A., Zaporozhets, I. A., Chenakin, S. P. *et al.* (1976) Orientation effects of secondary ion–ion emission from Mo single crystals. In: *Trudy VII Vsesojuznogo Sovesch. po fiz. vzaimodeistvija zarjazhenyh chastits s monokristallami.* Izd-vo MGU (in Russian).
30. Vasil'ev, M. A., Zaporozhets, I. A., Chenakin, S. P., and Cherepin, V. T. (1976). Single-crystal orientation effect on the secondary ion–ion emission In: *Vzaimodeistrvie atomnyh chastits s tverdym telom.* Izd-vo Khar'kov, Khar'kov, Un-ta (in Russian).
31. Lindhard, J. (1969) Slowing down of ions. *Proc. R. Soc. Lond.*, **A311**, 11–19.
32. Underdelinden, D. (1968) Single-crystal sputtering including the channelling phenomenon. *Can. J. Phys.* **46**, 769–777.
33. Bernheim, M. and Slodzian G. (1976) Crystalline transparency and anisotropy effect on backscattered noble gas ions. *Nucl. Instrum. Meth.* **132**, 615–621.
34. Bernheim, M. and Slodzian G. (1976) Lattice influence on ion emission under oxygen bombardment. *Int. J. Mass Spectrom. Ion Phys.* **20**, 295–304.
35. Bernheim, M. (1974) Influence des effects de reseau sur l'emission ionique secondaire incidence sur l'analyse quantitative et l'analyse en profondeur. Thesis, Centre d'Orsay.
36. MacDonald, R. J. (1974) An empirical relationship between atoms and ions sputtered from single-crystal surfaces. *Surf. Sci.* **43**, 653–656.
37. Blaise, G. (1976) Similarities in photon and ion emission induced by sputtering. *Surf. Sci.* **60**, 65–75.
38. MacDonald, R. J. and Martin, P. J. (1977) Quantitative surface analysis using ion-induced secondary ion and photon emission. *Surf. Sci.* **66**, 423–435.
39. Petrov, N. N. and Abrojan, I. A. (1977). *Surface diagnostics with ion beams.* Izd-vo LGU, Leningrad (in Russian).
40. Castaing, R. and Hennequin, J. F. (1966) Distirbution energetique et angulaire des ions secondaires emis par l'aluminium et le cuivre. *C. R. Acad. Sci. Paris* **262B**, 1008–1011.
41. Winograd, N. (1982) Characterization of solid and surface using ion beams and mass spectrometry. *Prog. Solid State Chem.* **13**, 285–375.
42. Cherepin, V. T., Kosyachkov, A. A., and Zaporozhets, I. A. (1984) *Angular resolved spectroscopy of low-energy ions.* Preprint IMF 10.84, pp. 1–44, Inst. Metal Phys. Acad. Sci. UkrSSR, Kiev (in Russian).
43. Cherepin, V. T., Kosyachkov, A. A., Dubinski, I. N., and Is'yanov, V. E. (1984) *Spectrometric complex with angular resolution for investigation of solid surfaces.* Preprint IMF 11.84, pp. 1–37, Inst. Metal Phys. Acad. Sci. UkrSSR, Kiev (in Russian).
44. Gibbs, R. A., Holland, S. P., Poley, K. E., Garrison, B. J. and Winograd, N. (1981) Image potential and ion trajectories in secondary ion mass spectrometry. *Phys. Rev.* **B24**, 6178–6181.
45. Rudat, M. A. and Morrison, G. H. (1975) Energy spectra of ions sputtered from elements by O_2^+: a comprehensive study. *Surf. Sci.* **82**, 549–576.
46. Arifov, I. A. (1968) *Interaction of atomic particles with a solid surface.* Nauka, Moscow (in Russian).
47. Düsterhöft, H. and Ihlenfeld, A. (1977) The energy distribution of positive secondary ions emitted from metal and semiconductor targets bombarded with 12 keV Ar^+ ions. – *Phys. Status Solidi* **A39**, K147–K150.
48. Vasil'ev, M. A., Krasjuk, A. D., and Cherepin, V. T. (1977) Mass spectrometer Mi–1305 with ion probe and energy analyzer for investigation of solids. *PTE*173–175 (in Russian).
49. Schubert, R. and Tracy, J. C. (1973) A simple, inexpensive SIMS apparatus. *Rev. Sci. Instrum.* **44**, 487–491.
50. Rouberol, J. M., Basseville, Ph., and Lenoir, J. P. (1972) Recent improvements of the ion analyzer and typical examples of applications. *J. Radioanal. Chem.* **12**, 59–69.
51. Lundquist, T. R. (1978) Energy distributions of sputtered copper neutrals and ions. *J. Vac. Sci. Technol.* **15**, 684.
52. Jurela, Z. (1973) Energy distribution of secondary ions from 15 polycrystalline targets. *Radiat. Eff.* **19**, 175–180.
53. Blaise, G. and Slodzian, G. (1973) Distribution energetique des ion secondaires. *Rev. Phys. Appl.* **8**, 105–111.
54. Gries, W. H. (1975) A formula for the secondary ion yield fraction emitted through an energy window. *Int. J. Mass Spectrom. Ion Phys.* **17**, 77–85.
55. Schroeer, J. M., Rhodin T. N., and Bradley, R. C. (1973) a quantum-mechanical model for the ionization and excitation of atoms during sputtering. *Surf. Sci.* **34**, 571–580.
56. Vasil'ev, M. A. (1979) Secondary ion emission at the bombardment of metal, alloy and

compound surfaces with ion beams. Doctoral thesis, Inst. metallofiziki AN USSR, Kiev. (in Russian).

57. Jurela, Z. (1975) Average energy of sputtered ions from 15 polycrystalline targets. *Int. J. Mass Spectrom. Ion Phys.* **18**, 101–110.

58. Slodzian, G. and Hennequin, J. F. (1966) Sur l'emission ionique secondaire des metaux en presence d'oxygene. *C. R. Acad. Sci. Paris* **263B**, 1246–1249.

59. Beske, H. E. (1964) Positive sekundärionenausbeute von 12 elementen. *Naturforsch.* **19a**, 1627–1632.

60. Ivashschenko, Yu. N. and Cherepin, V. T. (1970) Ion–ion emission from pure metals. *Dokl. Akad. Nauk Ukr, SSR.* A, **1970**, 141–143.

61. Vasil'ev, M. A., Chenakin, S. P., and Cherepin, V. T. (1974) Relative sensitivity of analysis performed by the ion–ion emission method. *Dokl. Akad. Nauk* SSR A, **1974**, 751–753.

62. Vasil'ev, M. A., Chenakin, S. P., and Cherepin, V. T. Determination of the coefficients of secondary ion relative yield (in Russian).

63. Fomenko, V. S. (1970) *Emissive properties of materials.* Naukova dumka, Kiev (in Russian).

64. Nemnonov, S. A. (1965) Electronic structure and some properties of transition metals and alloys of I, II and III large periods. *FMM* **19**, 550–568 (in Russian).

65. Sigmund, P. (1969) Theory of sputtering, I. Sputtering yield of amorphous and polycrystalline targets. *Phys. Rev.* **184**, 383–416.

66. Vasil'ev, M. A., Kosyachkov, A. A., Makeeva, I. N., and Cherepin V. T. (1982). Secondary ion emission from transition metals bombarded by hydrogen ions. *Phys. Met.* **4**(3), 515–524.

67. Cherepin, V. T., Kosyachkov, A. A., and Makeeva, I. N. (1984) Hydrogen ion bombardment in secondary ion mass spectrometry. In: *SIMS IV*, pp. 57–59, Springer, Berlin.

68. Benninghoven, A. (1967) Die positive Sekundärionenemission von sauerstoffbedeckten Metallen. *Z. Naturforsch.* **22a**, 841–846.

69. Tsai, I. C. C. and Morabito, J. M. (1974) The mechanism of simultaneous implantation and sputtering by high energy oxygen ions during secondary ion mass spectrometry (SIMS) analysis. *Surf. Sci.* **44**, 247–252.

70. Andersen, C. A. (1969) Progress in analytic methods for the ion microprobe mass analyzer. *Int. J. Mass Spectrom. Ion Phys.* **2**, 61–74.

71. Vasil'ev, M. A., Ivashchenko, Yu. N., and Cherepin, V. T. (1970) The effect of Fe–C alloy composition and structure on positive ion emission at ion bombardment sputtering. *Fazovye prevraschenija*, **32**, 143–148 (in Russian).

72. Kubashevsky, O. and Gopkins, B. E. (1965) *Oxidation of metals and alloys.* Metallurgija, Moscow (in Russian).

73. Kaminsky, M. (1967) Atom and ion collisions on metal surface. Mir, Moscow (in Russian).

74. Cherepin, V. T., Ivashchenko, Yu. N., and Vasil'ev, M. A. (1973) To the effect of ion bombardment on the Fe–C alloy corrosion resistance *Dokl. Akad. Nauk SSSR* **210**, 821 (in Russian).

75. Vasil'ev, M. A., Burmaka, L. S., Ivashchenko, Yu. N., *et al.* Investigation of iron surface after ion bombardment and oxidation in air with the use of AES method. In: *Vzaimo – deistvie atomnykh chastits s tverdym, telom*, Part 1, pp. 134–135, Naukova dumka, Kiev (in Russian).

76. Cherepin, V. T., Kosyachkov, A. A., and Vasil'ev, M. A. (1976) Effect of initial ion bombardment and oxidation on secondary ion emission. *Surf. Sci.* **58**, 609–612.

77. Vasil'ev, M. A., Kosyachkov, A. A., and Cherepin, V. T. (1976) The effect of Ar and the irradiation dose on iron oxidation (in Russian). *Dokl. Akad. Nauk* Ukr, SSR. A, **1976**, 267–269.

78. Vasil'ev, M. A., and Kosyachkov, A. A. (1976) Ion bombardment effect on alteration in Fe and Al work function. *Metallofizika* **65**, 108–110 (in Russian).

79. Benninghoven, A. (1975) Developments in secondary ion mass spectroscopy and applications to surface studies. *Surf. Sci.* **53**, 596–625.

80. Blaise, G. and Slodzian, G. (1968) Emission ionique des metaux de la premiere serie de transition. *C. R. Acad. Sci. Paris* **B266**, 1525–1528.

81. Fogel, J. H. (1972) Ion–ion emission — a new tool for mass spectrometric investigation of processes on the surface and in the bulk of solids. *Int. J. Mass Spectrom. Ion Phys.* **9**, 109–129.

82. Buhl, R. and Preisinger, A. (1975) Crystal structures and their secondary ion spectra. *Surf. Sci.* **47**, 344–357.

83. Vasil'ev, M. A., Kosyachkov, A. A., Nemoshkalenko, I. N., and Cherepin, V. T. (1979) Secondary emission of complex ions (review). *Metallofizika*, **1**, (in Russian).
84. Herzog, R. F. Poschenrieder, W. P., and Satkiewich, F. G. (1973) Observation of clusters in a sputtering ion source. *Radiat. Eff.* **18**, 199–205.
85. McHugh, I. A. and Sheffield, J. C. (1964) Secondary positive ion emission from aluminium surface. *J. Appl. Phys.* **45**, 512–15.
86. Morgan, A. E. and Werner, H. W. (1976) On the abundance of molecular ions in secondary ion mass spectrometry. *Appl. Phys.* **11**, 193–195.
87. Können, G. P., Tip, A., and De Vries, A. E. (1975) On the energy distribution of sputtered clusters. *Radiat. Eff.* **26**, 23–29.
88. Slodzian, G. (1975) Some problems encountered in secondary ion emission applied to elementary analysis. *Surf. Sci.* **48**, 161–186.
89. Staudenmaier, G. (1972) Clusters sputtered from tungsten. *Radiat. Eff.* **13**, 87–91.
90. Wittmaack, K. and Staudenmaier G. (1975) Diatomic versus atomic secondary ion emission. *Appl. Phys. Lett.* **27**, 318–320.
91. Morgan, A. E. and Werner, H. W. (1978) Molecular versus atomic secondary ion emission from solids. *J. Chem. Phys.* **68**, 15–20.
92. Salem, L. (1966) *Molecular orbital theory of conjugated systems*. W. A. Benjamin, New York.
93. Joyes, P. (1971) Alternations in the secondary emission of molecular ions from noble metals. *J. Phys. Chem. Solids*, **32**, 1269–1275.
94. Joyes, P. (1971) On a mechanism of secondary emission of polyatomic particles. *J. Phys.* **4**, L15–L18.
95. Joyes, P. (1972) Influence of asymmetrical correlations in the secondary emission of solid compounds. *J. Phys.* **C5**, 2192–2199.
96. Leleyter, M. and Joyes P. (1975) Secondary emission of molecular ions from light-element targets. *Radiat. Eff.* **26**, 105–110.
97. Vasil'ev, M. A. (1978) Theoretical models for the mechanism of secondary ion emission. *Metallofizika*, **72**, 3–11 (in Russian).
98. Rüdenauer, F. G. (1977) A comparison of quantitative models for SIMS analysis. *Mikrochim. Acta*, Suppl. 7, 85–94.
99. Williams, P. (1984) Current status of sputtered ion emission models. In: *Secondary ion mass spectrometry, SIMS IV*, p. 84, Springer, Berlin.
100. Joyes, P. (1969) Expulsion d'un electron lie due au choc de deux atomes d'un metal. *J. Phys. (Paris)* **30**, 243–251. Etude theoretique de l'emission ionique secondaire. *J. Phys. (Paris)* **30**, 365–376.
101. Blaise, G. and Slodzian, G. (1970) Processus de formation d'ions a partir d'atomes ejectes dans des etats slectroniques superexcites lors du bombardment ionique des metaux de transition. *J. Phys. (Paris)* **31**, 95–107.
102. Blaise G. and Slodzian, G. (1974) Evolution des rendements de l'emission ionique des alliages avec la nature an solute. I: Resultats experimentaux. *J. Phys. (Paris)* **35**, 237–241.
103. Sroubek, Z., Zavadil, J., Kubec, F., and Ždansky, K. (1978) Model of ionization of atoms sputtered from solids. *Surf. Sci.* **77**, 603–614.
104. Schroer, J. M., Rhodin, T. N., and Bradley, R. C. (1973) A quantum mechanical model for the ionization and excitation of atoms during sputtering. *Surf. Sci.* **34**, 571–580.
105. Gries, W. H. (1975) A formula for the secondary ion yield fraction emitted through an energy window. *Int. J. Mass Spectrom. Ion Phys.* **17**, 77–85.
106. Gries, W. H. and Rüdenauer, F. G. (1975) A quantitative model for the interpretation of secondary ion mass spectra of dilute alloys. *Int. J. Mass Spectrom. Ion Phys.* **18**, 111–127.
107. Sigmund, P. (1969) Theory of sputtering. I. Sputtering yield of amorphous and polycrystalline targets. *Phys. Rev.* **184**, 383–416.
108. Cini, M. (1976) A new theory of SIMS at metal surfaces. *Surf. Sci.* **54**, 71–78.
109. Antal, J. (1976) On the quantum theory of the emission of secondary ions. *Phys. Lett.* **A55**, 493–494.
110. Andersen, C. A. and Hinthorne, J. R. (1973) Thermodynamic approach to the quantitative interpretation of sputtered ion mass spectra. *Anal. Chem.* **45**, 1421–1438.
111. Werner, H. W. (1986) Quantitative secondary ion mass spectrometry. *Surf Interface Anal.* **2**, 56–74.
112. Winograd, N. (1982) Characterization of solids and surfaces using ion beams and mass spectrometry. *Prog. Solid State. Chem.* **13**, 285–375.
113. Harrison, D. E., Jr., Mason, W. A., and Webb, R. P. (1984) Molecular dynamics computer

simulation study of the damage produced in metal target surfaces during ion bombardment. In: *Secondary ion mass spectrometry, SIMS IV*, pp. 24–30, Springer, Berlin.

114. Sroubek, Z., Zavadil, J., Kubec, F., and Ždansky, K. (1978) Model of ionization of atoms sputtered from solids. *Surf. Sci.* **77**, 603.

115. Sroubek, Z., Ždansky, K., Zavadil, J. (1980) Ionization of atomic particles sputtered from solids. *Phys. Rev. Lett.* **45**, 580–583.

Chapter 2
SIMS Instrumentation

2.1. Ion sources

Any apparatus for secondary ion mass spectrometry includes a primary ion source, a vacuum chamber where the objects under study are placed, a mass analyzer, and a secondary ion detector.

The parameters of the primary ion source define to a high degree the characteristics of a given apparatus and the problems that may be solved. An ideal ion source must possess a high brightness, must produce an ion beam of homogeneous composition with a small energy spread, and with an ion current density that may be easily monitored and remains constant across the beam section. The source must be economical from the point of view of consumption of both energy and the working substance. It must operate at the lowest possible pressure in order to maintain good vacuum conditions in the sample chamber.

In many types of ion source the ions are extracted from plasma produced in an appropriate manner in a closed volume filled with the working substance (gas or vapour). The properties of the ion beam thus obtained are closely related to those of plasma itself and of the plasma boundaries where the ion extraction is performed, as well as to the properties of the extraction spacing where the primary beam is formed.

Operation of various ion sources is described in several monographs [1, 2] and reviews [3, 4]. However these works deal mainly with the sources intended for the use in various accelerating machines, where the magnitude of the ion current should be large while the beam diameter (in the SIMS sense) is of minor importance.

In the known laboratory and commercial secondary ion mass spectrometers, ion mass spectral microscopes, and ion probes, ion sources of various types are used. Most common are high-frequency sources [7–8], Penning-type sources, duoplasmatrons with hot and cold cathodes [10–21], surface ionization [2–24] and liquid metal ion sources [25–31], as well as sources of fast neutral atoms [23–33]. All the main types of ion source used in SIMS are shown in Figs 2.1 and 2.2 with unified symbols: the solid arrow refers to the gas flow, the dashed arrow to the ion beam; the light dash indicates magnetic

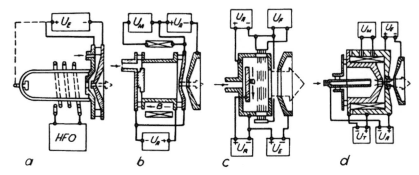

Figure 2.1 Types of ion sources.

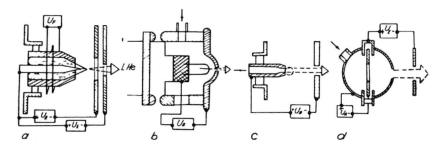

Figure 2.2 Types of ion sources.

materials, the bold dash non-magnetic materials. The power supplies are labelled as follows:

V_A = the anode (or discharge) voltage; V_E = the extraction potential; V_F = the filament potential; V_M = the power voltage for the magnetizing coil; V_R = the repeller voltage; HFO = the high frequency oscillator; V_T = the voltage on the intermediate electrode; B = the magnetic field direction.

2.1.1. The High frequency Source [Fig. 2.1(a)]
This is a simple and convenient device for the production of light gas ions. The working gas plasma is obtained from the electrodeless high-frequency discharge in the dielectric vessel (quartz or pyrex tube) where the vacuum (10^{-1}–1 Pa) is maintained. A high-frequency oscillator with a power of 50–200 W and a frequency 10–30 MHz produces a dense plasma from which the ion beam is extracted through a narrow orifice. Figure 2.1(a) shows two modes of how the extraction voltage can be applied. The disadvantage of such a source is the strong high-frequency interference, which hampers the functioning of the secondary ion detection system.

2.1.2. The Penning Source with a Cold Cathode [Fig. 2.1(b)]
This is also simple and highly reliable. Here the circular anode is placed between two cathodes, one of which has an orifice for the ion beam extraction. The whole system is inserted into a longitudinal magnetic field

which forces the electrons in the discharge to move on cycloidal orbits. Thus the working gas ionization efficiency becomes much higher but the working pressure is decreased. The use of permanent magnets and the optimization of electrode geometry allow the construction of a rather simple source with an operating pressure in the range 10^{-3}–10^{-2} Pa and with a discharge voltage which serves simultaneously as the extraction voltage. As a result, an ion beam dense enough for SIMS and possessing an energy of 4–4.5 keV is obtained using a single high-voltage power supply with an energy consumption of 4–5 W [20].

2.1.3. Source with Electron Impact [Fig. 2.1(c)]

This is well known and is widely used in mass spectrometry. Electrons from a heated cathode are directed by the field of a permanent magnet and pass through an orifice into the anode box. Electrons can be reflected by the anticathode in order to increase the efficient electron density. The gas pressure in the source is kept very low to avoid secondary ionization, and thus a linear relationship is created between pressure and the ion current at the exit. Moreover, since plasma is not produced, the extracted ion current may be increased with the aid of a repeller electrode placed in the anode box and held at a potential V_R. Ion currents in such a source are in the range 10^{-9}–10^{-8}A, but ions have a minimum energy spread, and the beam is of a high purity. Sources of this type are used in UHV systems with differential pumping for the study of surface processes.

2.1.4. Plasmatron Sources

Plasmatron sources provide the highest ion current density and the brightest beam of all ion plasma sources. The simplest version is the unoplasmatron, where the plasma density is increased by geometrical confinement within the conical body of an intermediate electrode. In this volume the plasma bubble is formed bounded to the double electrical layer. The bubble's outward surface has a spherical shape, which produces the focusing action for the plasma electrons at the exit orifice and thus ensures high ionization efficiency. The unoplasmatron is usually considered as the preliminary step on the way to the construction of a duoplasmatron, but the fact that the unoplasmatron is a simple source without magnets and can be used in a wide temperature range, makes it attractive for SIMS applications.

At present in various SIMS installations, use is often made of a duoplasmatron with hot and cold cathodes [Fig. 2.1(d)]. This source has some advantages over others e.g., a high intensity and very bright ion beam, good gas economy, and the possibility of obtaining positive, negative, and multiply charged ions [1, 2, 34–42].

In the duoplasmatron ions are extracted along the system axis from the plasma of a low pressure arc formed between a hot or cold cathode and anode. The exit orifice is bored in the anode along the discharge axis for the extraction of positive ions. In order to increase the plasma density and the ionization degree near the anode aperture, the discharge is confined by successive contracting action of the intermediate electrode and the strong axial magnetic field created by the magnetic lens, whose pole pieces are formed by the anode and intermediate electrode [Fig. 2.1(d)].

Negative ions can also be extracted from the duoplasmatron. For this purpose the discharge axis must be shifted about 0.5 mm with respect to the axis of the anode orifice, so that negative ions and electrons can be extracted from the peripherical part of the discharge. The electrons are then eliminated from the beam by an appropriate magnetic deflection.

2.1.5. The Liquid Metal Ion Source (LMIS) [Fig. 2.2(a)]

This consists of a low-volatility liquid metal film, flowing to the apex of a solid needle support structure with apex radius about 1–5μm. The application of an electric field of sufficient strength will deform the liquid film of the needle apex into a conical protrusion (Taylor cone) with the apex radius less then 50 nm. This cone forms in operation a sharper protrusion with still smaller radius so that very high field strength of about 20 V nm^{-1} is obtained. Metal ions are formed due to a field evaporation mechanism. Typically, an axial current intensity in excess of 20 μA sr^{-1} can be realized from LIMS in the 1–10 μA range of total current. If the virtual source size is about 50 nm then a source brightness of $1 \times 10^{10} A$ $m^{-2}sr^{-1}$ can be expected. The minimum beam energy spread of LMIS is about 5 eV and depends strongly on the total current and charge-to-mass ratio of ions [25]. Thus chromatic aberration limits the achievable beam size which at present can be obtained to less than 0.1 μm [26–31].

2.1.6. Gaseous Field Ionization Ion source [Fig. 2.2(b)]

This source gives a very bright hydrogen ion beam for submicron probe formation. The ions are field-desorbed from an emitter tip cooled close to liquid helium temperature. The tip is floated to high voltage to provide an accelerated ion beam. A sapphire block is used for thermal conduction between the tip mount and the cooling source — a liquid helium cold finger. The hole in the cap provides both the exit aperture for the beam and a differential pumping orifice. The source in the tip region is run at a hydrogen pressure of the order of 1 Pa, while the chamber containing the ion beam forming optics has a pressure of less than 1×10^{-3} Pa [24]. Special ion and fast neutral atom sources have been developed for fast atom bombardment SIMS (FAB). In these sources fast atoms are formed from ions by charge exchange processes. Two types of FAB sources are most frequently used: the capillaritron and the saddle field source.

The capillaritron ion gun [Fig. 2.2(c)]. This is small, rugged, and reliable; it produces a high-intensity narrow beam containing both ions and neutrals. The gas (Ne, Ar, Xe, etc.) is supplied to a hollow jet. Potential is applied to the jet (anode) with respect to the cathode, which is usually connected to earth. As the gas atoms emerge from the jet orifice they form a very localized region of high pressure and are ionized by electrons traversing the inter-electrode space. The resulting ions are accelerated through the extractor cathode as a narrow beam. the beam energy makes up a few eV of the potential applied to the jet (anode) [33].

About 2% of the gas is ionized. The remaining gas close to the jet orifice and travelling with only thermal energy provides conditions for the resonance charge transfer. This results in a narrow distribution of both the ions and neutrals centred along the beam axis. Conversion of input energy into the output beam amounts to almost 100%, so that the gun remains cold.

The saddle field source [Fig. 2.2(d)]. Electrons are forced to oscillate between two cathodes under the action of a d.c. field. Electrons originating from a sector of the cathode travel through the anode region towards the opposite cathode sector. They are retarded, return, and continue to oscillate around the central saddle point in the potential field. The probability of the working gas ionizing is high because of long electron paths, so that a discharge can be maintained at relatively low pressure [32]. Positive ions formed in the discharge travel radially towards the cathode and emerge from the orifice. The mechanism of beam formation in the saddle field source is such that the internal field energizes the ionizing electrons and aligns the ions in such a way that most of them are neutralized by electrons within the source, thus giving rise to the beam of neutrals. The composition of the beam can be markedly affected by an external field. When an electrode of about 300 V is situated 2 cm from the cathode, the neutral content of the beam can be reduced to less than 5%. Conversely, the ion content can be reduced to less than 1% by careful shielding of the cathode region. The ion energy spread for several typical sources is shown in Fig. 2.3.

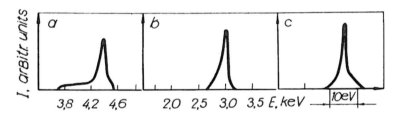

Figure 2.3 Energy spread of ions obtained from different sources.
(a) High frequency source, capacitance coupling, oscillator frequency 220 MHz, extracting voltage 4.5 kV..
(b) Penning source, extracting voltage 3.0 kV.
(c) Duoplasmatron [5].

2.2. Mass separation of the primary beam

When an ion beam produced by the ion source contains undesirable ion species it is necessary to introduce a mass separator into the beam column. This is the case when a noble gas primary beam is obtained from a plasma source. Impurities, always present in such a beam, induce uncontrollable chemical enhancement effects in SIMS surface investigations. A liquid metal ion source fed with eutectic alloys, or even pure metals, also requires mass separation.

Various types of mass separator with homogeneous and inhomogeneous magnetic 90 and 180° fields are available [43], but the Wien filter ($E \times B$ − filter) is the most suitable for this purpose [44, 45]. This type of separator has two advantages. First, it has a straight axis and hence is mechanically simpler, and second, it can be built with a permanent magnetic field and still allow passage of different selected ions with a given energy.

For ions of eV energy and M mass number an effective radius r_0 can be defined from

$$r_0 = 144 \, \frac{\sqrt{MV}}{B_0} = 2 \, \frac{V}{E_0} \qquad (2.1)$$

where r_0 is in cm, V in V, B_0 in Gauss, and E_0 in Vcm^{-1}. r_0 is the radius of curvature of the trajectory which ions would follow in the magnetic field B_0 without an electric field or in the electric field E_0 without a magnetic field. This effective radius and the length L of the filter define the focusing and dispersive properties.

The Wien filter with uniform fields has focusing action in the direction of dispersion only. Stigmatic focusing can be achieved by making one of the two fields non-uniform [44], so that either

$$B_x = B_0 \left(1 - \frac{x}{2r_0} \right) , \, B_y = - B_0 \frac{y}{2r_0} \qquad (2.2)$$

or

$$E_x = E_0 \left(1 + \frac{x}{2r_0} \right) , \, E_y = - E_0 \frac{y}{2r_0} \qquad (2.3)$$

This is the case when either magnetic fields lines or the equipotential surface through the axis have the radius of curvature $2r_0$ (Fig. 2.4).

The Wien filter may be regarded as a thick lens with two principal planes (Fig. 2.5). The same two principal planes may be used to describe dispersive action. The trajectory of an ion entering at an angle α_1 to the axis forms the angle α_2 with the axis behind the field, and

$$\alpha_2 - \alpha_1 = \gamma \left(\frac{\Delta M}{M} - \frac{\Delta V}{V} \right) - \frac{x_1}{f} \qquad (2.4)$$

The first term in this expression describes the dispersive action while the second term — the focusing action.

The dispersive factor ν and the focal length f are listed in Table 2.1. Note that mass dispersion and energy dispersion are oppositely directed, in contrast to a magnetic sector field. The high energy tail of a mass peak in a mass spectrum taken with the Wien filter is therefore on lower mass side [42].

From Equation (2.1) one obtains:

$$M - V \left(\frac{B_0}{72 \, E_0} \right)^2 \qquad (2.5)$$

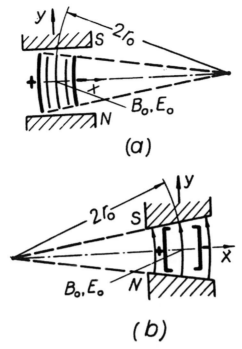

(a)

(b)

Figure 2.4 Stigmatically focusing Wien filter. (a) Uniform magnetic field B_0, non-uniform electric field. (b) Non-uniform magnetic field, uniform electric field, E_0.

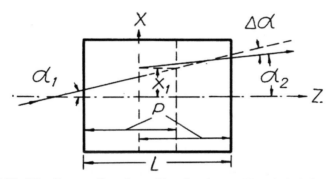

Figure 2.5 The Wien filter as a dispersive and focusing element. The principal planes are located at a distance p from the boundaries [47].

With the beam energy eV being fixed and with constant electric field strength E_0 the mass spectrum can be scanned by varying the magnetic field strength B_0. In this case r_0 remains constant and consequently the same is true for the dispersive and focusing parameters in Table 2.1. Only in this mode is it possible to apply one of two ways of constructing the stigmatically focusing Wien filter that are indicated in Fig. 2.4.

In the other mode, where the electric field E_0 varies while B_0 remains constant, r_0 varies with E_0 according to Equation (2.1) and so do the

Table 2.1
Parameters of Wien filter

Parameters	Uniform fields	Stigmatic imaging	$r_0 \gg L$	
			Uniform	Stigmatic
p	$r_0 \tan \dfrac{L}{2r_0}$	$\sqrt{2}r_0 \tan \dfrac{L}{2\sqrt{2}r_0}$	$\dfrac{L}{2}$	$\dfrac{L}{2}$
f	$r_0 \dfrac{1}{\sin(L/r_0)}$	$\sqrt{2}r_0 \dfrac{1}{\sin(L/\sqrt{2}r_0)}$	$\dfrac{r_0^2}{L}$	$2\dfrac{r_0^2}{L}$
ν	$\dfrac{1}{2} \sin \dfrac{L}{r_0}$	$\dfrac{1}{\sqrt{2}} \sin \dfrac{L}{\sqrt{2}r_0}$	$\dfrac{L}{2r_0}$	$\dfrac{L}{2r_0}$

p – distance of principal planes; f – focal length; ν – dispersion factor; L – length; r_0 – effective radius (from [47]).

dispersive and focusing parameters in Table 2.1. Hence, the dispersive as well as the focusing actions are stronger for the light masses than for heavier ones. This must be kept in mind when the Wien filter with permanent magnet is to be incorporated in the beam column.

An effective method to achieve stigmatic focusing with the permanent magnetic field is to make the magnetic field uniform and fit the plane electric field plates into the gap with potentials adjustable against the magnet pole pieces so that the latter act like Matsuda plates [46] and allow the curvature of the median equipotential surface to be adjusted. In this case conditions are mostly such that $r_0 \gg L$ (Table 2.1), and the filter can be regarded as a thin lens positioned in the middle of the length L, the focal length of the stigmatic case being twice that of the uniform field case.

A very compact Wien filter can be constructed with samarium–cobalt permanent magnets. Liebl [47] constructed a filter which fits vacuum tubing 10 cm in diameter. It is only 8.6 cm long, has a field strength of 0.4 T across the 1 cm gap, and is capable, in combination with an einzel lens, of unit mass resolution for masses and energies commonly used in SIMS.

2.3. Primary and secondary ion optics

In a typical SIMS apparatus the sample is bombarded with a fine beam of ions. Secondary ions that are emitted from the sample are collected, focussed, and their energy is analyzed before they are injected into a mass filter. All these procedures need some kind of ion optics.

In order to produce a fine spot carrying as much current as possible, the ion source is usually followed by lenses (typically two) which demagnify the source. On the other hand, if the source size is very small, which might be the case for a metal ion source, it might be necessary to magnify it. Electrostatic lenses are generally more efficient for focusing low energy ions than magnetic lenses. The most commonly used lenses in ion optics are two-electrode lenses, the immersion lens, a whole multitude of three electrode lenses, and a few multielectrode designs [48].

Immersion lenses often form the basis for primary ion source geometry and secondary ion collection optics, while three-electrode rotationally symmetric lenses (einzel lens) are most frequently used for focusing purposes.

In microbeam designs, einzel lenses can be run with earthed spaces between similar lenses to enable other optically active elements to be placed in the spaces. These other elements include deflection and alignment plates and stigmators [13, 14].

The einzel lens can be run in a retarding or accelerating mode, i.e. with the central electrode retarding the ions as they enter the lens and accelerating them as they leave, or accelerating ions in and retarding them out. The accelerating type of lens usually has a lower spherical aberration coefficient for a given working distance or focal length [49]. However, the voltage applied to the middle electrode has to be much higher than the voltage of opposite polarity applied to the lens working in the retarding mode when it gives a similar focal length. The disadvantages associated with the higher voltage generally outweigh the advantages of the accelerating lens and so in probe systems and microscope applications retarding lenses are more often used.

The properties of various lens systems have been tabulated by various authors, and comprehensive data on two-aperture, three-aperture, and two-tube lenses are available in the literature [50–53].

In terms of practical details the following factors should be taken into account when designing an ion optical system. First of all, the inter-electrode spacings must be sufficiently large to avoid breakdown; suitable tracking distances must be provided along the insulators. For polished electrode surfaces the spacing between electrodes should be more than approximately 1 mm per 10 kV for electrons and 1 mm per 6 kV for ions. The tracking distance on insulators should be more than 1 mm per kV. The inter-electrode spacings are larger for ion optics than for electron optics because when ions interact with a metal surface they produce many secondary electrons and hence increase the probability of an electrical breakdown, in addition to the ion sputter of neutral atoms, which can be deposited on insulators as conducting film.

Suitable materials for electrodes are non-magnetic stainless steels and some non-magnetic aluminium alloys; sometimes copper and titanium can also be used. Magnetic screening for ions is practically unnecessary.

Care must be taken that the incoming ions and the ion beam have no direct 'viewing' of any insulator to avoid charging up, which can cause astigmatism, deflection, and instabilities in the beam current.

One more important aspect of electrostatic lenses is the effect of mechanical tolerances (i.e. circularity of bores, accuracy of alignment, etc.). If the lens bores are non-circular or are displaced laterally with respect to each other, then astigmatism is introduced into the optical system. Practical experience shows that in high-resolution ion probes, the ellipticity of the final lens bore must not be significantly greater than the spatial resolution expected of the system [48].

2.4. Mass spectrometers

Secondary ions can be mass analysed by means of mass spectrometers with a sector magnetic field of any type. In such spectrometers the mass scanning is performed with a constant accelerating voltage, and mass spectral resolution must be not worse than 300. It should be mentioned that it is not a simple task to obtain such a resolution, since the energy spread of secondary ions is rather large. Therefore in studies that require measurements of heavy ions it is necessary to use a double-focusing mass spectrometer, of Mattauch–Herzog type for example. It is possible to use mass analyzers which, owing to peculiarities in their operation, are less sensitive to secondary ion energy spread. Quadrupole and monopole structures may be used as such analyzers.

Special experimental setups for SIMS are as a rule rather complex pieces of apparatus [7, 10, 21, 55–61], but many problems may be solved by using commercial magnetic mass spectrometers equipped with a secondary ion source. A simple and inexpensive mass spectrometer such as the MI-1305 is the basis for the development of several SIMS instruments [62].

In the majority of practical cases it is sufficient for SIMS to use mass analyzers with a resolution $M/\Delta M = 300$–400. However, when complex or cluster ions with a large mass are studied, instruments with a much higher resolution are required in order to resolve possible overlapping of different ion species. Table 2.2 summarizes the resolutions necessary to resolve elemental ions with mass numbers 27–238 from hydrocarbon ions, as well as multiply charged ions from polyatomic ions in the range of mass numbers 14–58. High-resolution SIMS is usually accomplished using classical schemes of double focusing Mattauch–Herzog mass spectrometers [15, 16, 57, 58]. Construction of such systems is, strictly speaking, reduced to the introduction of a primary ion source into spark mass spectrographs, along with corresponding extraction optics to transfer secondary ions to the entrance slit of mass spectrograph. Since the latter process is extremely complex, its use is justified only when the problem cannot be solved in simpler ways. The last decade and a half have witnessed a wide application of quadrupole and monopole mass filter systems [63–66] for SIMS. They are relatively simple in construction and can be used with UHV. The energy spread of the secondary ions has a comparatively weak effect on the mass-spectral resolution. These specific features make such filters rather attractive for use in investigations of

Table 2.2
Resolution (at 10% peak height) necessary for separation of elemental, hydrocarbon, multiply charged and polyatomic ions

Mass number	Doublet	$M/\Delta M$	Mass number	Doublet	$M/\Delta M$
27	$C_2H_3^+ - {}^{27}Al^+$	642	14	$Si^{2+} - N^+$	960
63	$C_5H_3^+ - {}^{63}Cu^+$	670	28	$Si^+ - Fe^{2+}$	2980
138	$C_{10}H_{12}^+ - {}^{138}Ba^+$	570	56	$Si_2^+ - Fe^+$	2895
208	$C_{15}H_{28}^+ - {}^{208}Pb^+$	866	58	${}^{28}Si{}^{30}Si - Ni^+$	3800
238	$C_{17}H_{34}^+ - {}^{238}U^+$	1100			

SIE processes. Liebl [66] has pointed out that the first quadrupole for SIMS was built by Krohn in 1961 [67]. in 1967 it was announced that in Lyon a micro-analyzer with quadrupole mass filter had been built. Later the quadrupole was used by McDonald in Canberra for SIE analysis. In 1977 Benninghoven and Loebach built an instrument with quadrupole for SIMS in UHV [68]. In 1972 we described a secondary ion mass spectrometer with a monopole mass filter [9]. Since then Wittmaack *et al.* have constructed DIDA [69], and in 1973 Schubert and Tracy proposed an instrument consisting of a quadrupole and an energy analyser with a half of a cylindrical mirror placed in front of the quadrupole [70]. These works were followed by a number of publications on the use of quadrupole mass filters in experimental and commercial instruments (see references in [66]). At present this type of analyser may be without exaggeration considered as a leading one in the applied SIMS [71]. An important advantage of quadrupole lies in the possibility of setting the resolution by purely electrical means without mechanical adjustments. Another attractive feature is the linear dependence between the mass number and the amplitude of the applied d.c. or a.c. voltage, since this allows easy identification in the mass scale. The absence of inertial elements in the quadrupole makes it possible to realize different versions of fast scanning of the mass scale, selection of relevant peaks in any succession, etc. In real instruments the quadrupole rods are of a finite length, this being of importance for the ultimate possible resolution, which amounts to $R = n^2/12.2$, where n is the number of alternating voltage cycles during the time when an ion is moving in the quadrupole. Hence for adequate resolutions it is necessary to use a high operating frequency, to use sufficiently long rods and to decrease the energy of ions that are analysed.

In modern quadrupoles with high resolution (up to some thousands $M/\Delta M$) and a wide mass range (1000 or more) the rods are 250–300 mm long, the frequency is 1–5 MHz, the ultimate voltage is 3–5 kV and the power of high frequency generators makes up several hundred watts.

Since ions of different masses possess almost the same energy, E_z in the z-direction, we get

$$n = fL(m/2E_z e)^{1/2} \qquad (2.6)$$

where f is the frequency of the alternating component, L is the rod length, and m and e are the ion mass and the charge, respectively. It may be seen that the maximum attainable resolution grows in proportion to the ion mass. As a result, the mass spectrum recorded with a constant scanning rate has equidistant peaks throughout the mass scale. These instruments are characterized by the resolution R, which refers to the mass number M as

$$R = KM \qquad (2.7)$$

For instruments of moderate quality $K = 1.0–1.2$, while for those of high quality $K = 5–10$.

It must be noted that operation and parameters of quadrupole and monopole filters depend on the accuracy of the rod manufacture and positioning (to micrometer tolerance) as well as on the stability of the applied voltage (whose fluctuations should not exceed $100\ R^{-1}$).

For SIMS instruments with a quadrupole mass analyzer the parameters of the secondary ion beam must be consistent with quadrupole input parameters. Although the secondary ion energy spread has only a minor effect, the beam must nevertheless be energy-filtered, because its high-energy 'tail' extends to hundreds of eV. This 'tail' is too large for the high-resolution quadrupole. On the other hand, the secondary ion energy distribution has its maximum at 10–20 eV, that is ions may be admitted into the quadrupole directly at their initial energies without any acceleration. This, of course, decreases the efficiency of secondary ion collection and the analysis sensitivity, but many authors prefer simple ion optics and often go no further than to place a simple energy filter in front of the quadrupole (Fig. 2.6).

The commercial differential in-depth analyser (DIDA) is a typical instrument for use in SIMS with a quadrupole [54, 69, 47] (Fig. 2.7). Here the high-energy secondary ions are filtered with a simple parallel plate capacitor to increase the resolution of a mass filter. Since the quadrupole has the axial symmetry it is perfectly suited for the analysis of beams possessing a circular cross-section. Most energy analyzers of deflecting type produce at their exit a more or less flat beam, resulting in additional loss of ions. Much more efficient are energy analyzers with axial symmetry. In one of them [60] the energy analysis is performed through a double deflection by electrodes of a conical shape (Fig. 2.8). In another case [73, 74] the system of hemispherical grids placed between a sample and the quadrupole serves for energy filtering (Fig. 2.9). The grid potentials are selected in such a way that secondary ions

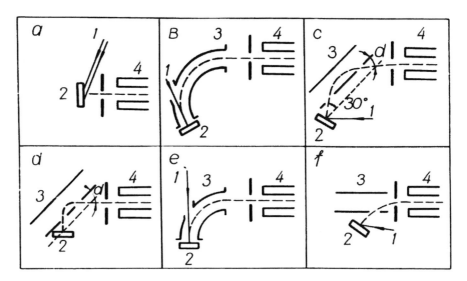

Figure 2.6 Input energy analyzer in SIMS with quadrupole. (a) Direct ion introduction. (b) 127°-capacitor. (c) Half-cylindrical mirror. (d) 90°-parallel plate mirror. (e) Cylindrical capacitor. (f) Parallel plate capacitor. (1) Primary beam; (2) target; (3) energy analyzer; (4) quadrupole.

are at first decelerated and then accelerated. Only ions of a certain energy are focused on the quadrupole entrance aperture; all the others are filtered out.

It should be mentioned that a direct viewing between the sample surface and the detector must always be avoided, since concomitant processes which occur during ion bombardment lead to creation of fast neutral particles, photons, and electrons. When the latter penetrate into the quadrupole and then into the ion detector, they cause background noise, and thus sharply decrease sensitivity.

It is worth noting that in case of a pulse ion-bombardment, time-of-flight mass spectrometers (see e.g. [75–77]) are well-suited for SIMS.

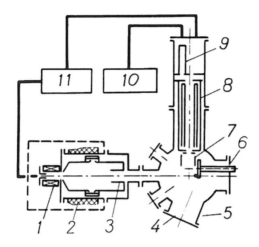

Figure 2.7 Diagram of DIDA. (1) Ion source; (2) insulator; (3) entrance aperture; (4) viewing port; (5) vacuum chamber; (6) sample holder; (7) parallel plate capacitor; (8) quadrupole; (9) chenneltron; (10) recorder; (11) power supply.

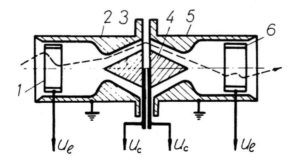

Figure 2.8 Deflecting energy analyser with axial symmetry. (1) Collecting lens; (2) leading shield; (3) leading cones; (4) output cone; (5) output shield; (6) output lens.

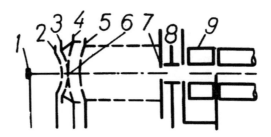

Figure 2.9 Retarding–accelerating energy analyser for SIMS. (1) target; (2, 3, 4, 5,) grids; (6) central stop; (7) exit aperture of the analyser; (8) lens; (9) quadrupole.

2.5. Ion detection and vacuum systems

In all SIMS instruments use is made of both the usual ion detectors (simple collectors, Faraday cages, SEM and channeltrons) and specific systems developed for secondary ion detection and efficient measurement of positive and negative ion currents (with the possibility of cutting off the high-energy ions).

The simplest ion detector with energy analysis was designed in our laboratory and is intended for use with a standard commercial magnetic mass spectrometer [78].

When SEM and CEM are used to count ions one must keep in mind that fast neutral particles, photons, and electrons, when they reach the multiplier entrance, produce strong background noise, thus narrowing the dynamic range. Therefore it is a common practice to place SEM and CEM in an off-axis position with respect to the quadrupole or monopole to avoid direct viewing. Ions are deflected by corresponding fields (see e.g. [79]). The background suppresion may be achieved by modulation of the primary ion beam. Alternatively a lock-in amplifier and synchronous detector in the secondary ion detection channel may be used.

For SIMS in a wide mass range account should be taken of the mass discriminative effect in SEM and CEM with increasing ions mass. The number of ions being the same at the detector input, the signal at its output is smaller for heavy ions. Systematic study has shown [80] that the effect of ion mass m_i on the detection efficiency γ_i is satisfactorily described as

$$\gamma_i = C m_i^{-0.4} \tag{2.8}$$

where C is the constant in the mass range from $^{14}N^+$ to $^{238}U^+$. Discriminative effects become weaker with energy growth at which ions impinge the surface of the first dynode in SEM or CEM [81]. Therefore in some instruments the ion detectors are used with a preliminary conversion of the ion current into the corresponding secondary electron current. To achieve this the metallic electrode is bombarded under a high potential (30–40 kV) with the ions to be detected. Secondary electrons produced in this way are accelerated and directed to scintillator, the light output from which is registered by PEM situated outside the vacuum system. Such a detector was first proposed by

Daly [82]. It was later improved and is now applied in some experimental installations [83–86].

SIE as a surface process is very sensitive to the effect of adsorption of residual gas molecules in the vacuum chamber. Therefore severe requirements are imposed on the vacuum system.

A high background of hydrocarbon peaks in secondary ion mass spectra results mainly from the fact that the residual gas atmosphere in the vacuum system can be polluted by oil vapours when oil diffusion pumps are used. hence it is advisable to use oilless vacuum systems for SIMS, with mercury-vapour, cryosorption, turbomolecular, and ion pumping facilities.

Working gas pressure in the discharge region of ion source is kept at 10–10^{-2}Pa, while in the anlytical chamber, where samples are bombarded, vacuum must be not worse than 10^{-4}–10^{-5}Pa. Here the differential pumping with the aid of an additional vacuum system may be rather helpful.

In order to maintain identical vacuum conditions when investigating different objects an air-lock system must be used, which should provide fast sample introduction and removal. Air-lock valves in UHV systems have metallic seals; samples are transferred into the chamber and out from it by means of a transporting rod which moves in a sufficiently long tube made of stainless steel. The sample holder is attached to one end of the rod, while the other end has a ferromagnetic termination. A strong permanent magnet moving outside the tube shifts the ferromagnetic termination along and thus controls sample position. In order to obtain reliable data with high reproducibility, especially when investigating the effect of different treatments on SIE from targets of the same chemical composition, a set of samples is simultaneously introduced into the analytical chamber and analysed in succession without any alteration in the vacuum or other experimental conditions.

Very severe vacuum requirements are established for the analysis of microimpurities or surface phenomena [17]. In this case all the system components must be machined from stainless steel and heavy monoisotopic metals, in particular from tantalum.

2.6. Mass – spectral microscopes

Secondary ions are produced at those points of a bombarded surface where the impinging ion transfers its energy to surface atoms in the sample crystal lattice. Therefore secondary ions represent atoms of a definite kind localized in particular micro-regions of the surface being analysed.

If the number of secondary ions ejected from micro-regions is known, one may estimate the initial quantity of corresponding atoms, that is, define their local concentration. In other words, the measurement of a local density of secondary ion current may under certain conditions serve as the basis for the analysis of local chemical or isotopic composition of solid surfaces. Along with this, the ion sputtering of the surface is a process connected with ion penetration into material, so that deeper layers are revealed. Successive analysis of secondary ion local distribution during sputtering offers the unique possibility to study three-dimensional concentration distribution.

There are three different types of instrument designed for local measurements of secondary ion currents in selected areas of the surface under study.
In the first type of instrument [Fig. 2.10(a)] secondary ions ejected from the
sample surface are accelerated and focused by the emission lens, which
produces a magnified ion image. This image is mass-separated by a stigmatic
mass analyzer with adequate optical properties. Ion images separated by the
analyzer and containing ions of only certain type are converted into electron
images, and the latter are transformed into the optical images, which are
observed on a fluorescent screen and recorded on photographic film. Such
installations have the name of ion mass spectral microscopes.

Instruments of the second type have a construction such that the primary
ion beam irradiates a minimal area on the target surface. The diameter of this
area determines the attainable locality of the measurement and spatial
resolution. Qualitative and quantitative composition analysis of secondary
ions emitted from the irradiated area is carried out by a mass spectrometer.
The use of a television technique for primary beam scanning synchronously
with the beam of CRT, the brightness of which is modulated by the output
signal from the mass spectrometer, makes it possible to obtain a magnified
image of the sample surface representing the distribution of given ion species.
Instruments of this type are known as ion microprobes [Fig. 2.10(b)].

In instruments of the third type the magnified ion image of the irradiated
surface is produced by the emission lens. Then by means of a dissector
aperture a small area is selected in this image, and ions from this selected area
are mass-analysed. Alternatively the ion image can be scanned along the
aperture, the mass analyzer being set for a certain ion species. The output
signal from the mass spectrometer modulates the CRT brightness at synchronous scanning [Fig. 2.10(c)]. Such an instrument may be defined as a dissector
mass spectral microscope.

Mass-spectral microscopy is a comparatively young technique. The first

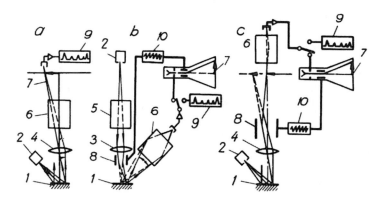

Figure 2.10 Types of instruments for local SIMS analysis. (a) Ion microscope. (b) Ion
microprobe. (c) Ion dissector microscope. (1) Sample; (2) ion source; (3) primary ion beam
optics; (4) secondary ion optics; (5) mass filter of primary ions; (6) mass filter of secondary ions;
(7) mass filtered ion image; (8) deflecting system; (9) mass spectrum recorder; (10) scan
generator.

mass-spectral microscope was built by Castaing and Slodzian at the beginning of 60s]7, 87]; the first ion microprobe was constructed by Liebl in 1967 [11]; and the first dissector microscope was developed in our laboratory in 1977 [88].

A block diagram of one of the simplest mass-spectral microscopes is shown in Fig. 2.11. This microscope was built in the Institute of Metal Physics at the Academy of Science, UkrSSR (Kiev), in co-operation with the Physico-Technical Institute of Low Temperatures [89–93]. Its main features reproduce the ion-optical principles realized for the first time by Slodzian [7]. The primary ion beam is created by a high-frequency plasma source. The working gas plasma is produced in a Pyrex tube (2) as a result of a high-frequency discharge in a vacuum of about 10^{-1} Pa. The high-frequency voltage from the oscillator (50 mHz, 40 W) is applied to the rings (3). Ions are extracted from the plasma when the extracting voltage (about 2 keV) is applied between the anode (1) and the probe (4). Ions are accelerated to 10 keV in the immersion lens (5) and then are focused by the condensor lens (6) on the sample surface. The bombardment angle is set by the deflection plate (7). Interaction between the ion beam and the sample (8) [which is the cathode of the emission objective (9) and has a potential of the order of +3.5–4.0 kV] results in secondary ion emission, which displays the sample surface composition. These ions are accelerated and focused by the emission lens, so that a real ion image, magnified 10–20 times, is produced as a superposition of elemental images formed by different ion species. Chromatic aberrations are reduced by the contrast aperture (10) placed in the crossover of the emission lens. This aperture cuts off ions having a high tangential velocity component. The ion beam transferring the ion image of the surface passes through the mass analyzer (11), where the image formed by ions of a definite mass-to-charge ratio is selected. The mass analyzer (11) provides stigmatic imaging due to the focusing action of fringing fields with oblique incidence and exit of the ion beam. The tilt angle may be adjusted by rotating terminations of the magnetic prism pole pieces (12). The residual astigmatism of the image after mass separation is corrected by the octopole electrostatic stigmator (13).

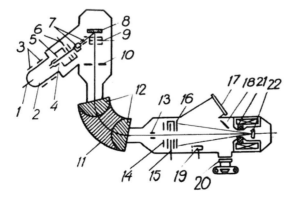

Figure 2.11 Diagram of simple ion mass spectral microscope (for details see text).

An elementary image selected by means of the selection slit (14) is projected by the teleobjective (15) on the cathode of the image converter (22), having a negative potential of about 10 kV. Secondary electrons knocked out from the cathode at ion bombardment are accelerated by a two-electrode immersion lens whose anode (21) serves as a pole piece of the magnetic lens [94]. The latter does not affect the ion image but provides the focusing of electron image observed on the screen (16) through the viewing port (17). With the aid of the mirror (18) placed at an angle of 45°, the image is photographed via the viewing port by the camera (20), situated outside the microscope chamber.

The instrument described serves not only as an ion microscope but also as a mass spectrometer for in-depth profiling. In the latter case the filtered ion beam is collected by Faraday cage (18), the ion current is amplified by the picoammeter, and the mass spectrum is recorded.

The most important components of any mass-spectral microscope are the emission objective, the magnetic image separator, and the image converter (which converts the ion image into an optical one). Different versions of these components are described in detail in [7, 87–99].

Theory and methods of engineering calculations are by now well enough developed, and ion-optical systems have already been constructed where mass-spectral resolution of the order of some thousands $M/\Delta M$ is achieved at an optical resolution better than 1 μm. The top achievement in this field is the IMS–3F CAMECA microscope, based on Slodzian's ideas and findings (see diagram in Fig. 2.12). The primary beam is produced by a duoplasmatron with a cold cathode (3). A three-lens condenser system (2) forms an ion spot 3–500 μm in diameter. Secondary ion optics include the emission lens and the transfer optics, which consist of three lenses to provide crossovers of various

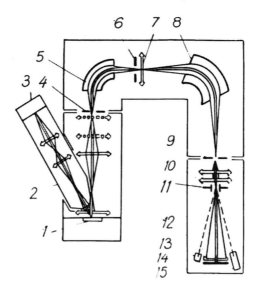

Figure 2.12 Ion-optical diagram of microanalyser microscope IMS–3F. (For details see text).

size, that is various magnifications, in the plane of the spectrometer entrance aperture (4). A stigmatic toroidal energy analyzer (5) allows filtering of a certain energy band of secondary ions, this band being limited by the selector slit (6). The image transfer and its focusing in the achromatic plane of the magnetic prism is performed by the lens (7). The stigmatic magnetic prism (8) ensures mass and energy focusing for ions of a definite mass-to-charge ratio on the exit slit (9), where a virtual image is obtained. The latter is converted into a real one by the double electrostatic projection lens (10). The deflection system (11) directs this image onto the microchannel plate (12) followed by the fluorescent screen (14). Alternatively, the image may be directed onto the ion receiver (the Faraday cage) (13) or the SEM (15).

The use of transfer optics and a double focusing system allow the IMS–3F to achieve mass-spectral resolutions of 5000 in the mass spectrometer mode and up to 10 000 in the mass spectrograph mode. In the latter case the exit crossover (slit 9) is projected onto the microchannel plate instead of the sample surface, so that a part of the mass spectrum is imaged with a high mass resolution. Variation of the entrance crossover size by the transfer optics in such a way that the size corresponds to entrance slit dimensions allows increased sensitivity and mass-spectral resolution, but at the expense of the area size from which secondary ions are collected. In this way a good lateral resolution is combined with a high mass resolution.

Mass spectral microscopes of the above type have some disadvantages. First of all, the ion-optical components are complex, because the mass analyser participates in ion image formation. Also, it is difficult to analyze the same spot on the sample surface for different elements, or to perform a local in-depth analysis for various elements, since the image on the screen is shifted when the spectrometer is retuned from one mass to another.

We have designed a dissector ion microscope–microanalyser where the properties of the ion microprobe are combined with those of a mass spectral microscope so that new possibilities arise for different analyses of solid surface and bulk material [88, 100]. The ion-optical scheme of this microscope is shown in Fig. 2.13. Secondary ions ejected from the sample (1) are accelerated and focused by the emission objective (2), which has a contrast aperture (3). The ion image can be deflected in two mutually perpendicular directions with the aid of the deflecting system (4). Then the image can be projected, magnified or demagnified by the transfer optics (5). An ion–electron image converter of known type [101] comprises the anode (7), vehnelt (8) and cathode (9). A new element in this converter is the dissector aperture in the cathode. Depending on the size of this aperture and the ion image magnification at the surface of the cathode (9), the aperture transmits the ion current from a limited area on the sample surface, thus ensuring the spatial resolution. Ions accelerated in the converter field knock secondary electrons out of the cathode, and the electron image is formed as a result. A small magnetic prism (6) deflects secondary electrons to the fluorescent screen (18) where they can be observed. The ion–optical converter is followed by a decelerating system composed of two lenses formed by tubular electrodes (10, 11), the first electrode being mechanically and electrically connected

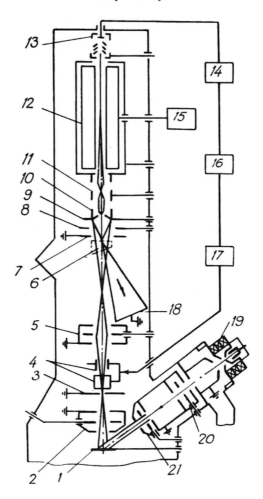

Figure 2.13 Dissector ion microscope. (14) ion detection system; (15) quadrupole generator; (16) recording and display; (17) scan generator. For other details see text.

with the cathode of the ion–electron converter, while the second is connected with the shield of the quadrupole mass filter (12), which is kept under a high potential to ensure that the ion beam has enough energy (after it has been decelerated) for quadrupole operation. The energy of the ions that enter the mass filter is determined by the difference of potentials between the sample and the shield, and may be easily controlled. The quadrupole is followed by the ion detector (13), which includes the exit accelerating lens, the deflecting system, and the SEM. The system designed to generate and shape the primary ion beam includes the duoplasmatron with a cold cathode (19); a focusing system composed of two condenser lenses (20, 21) placed in series; and deflecting plates (not shown) for primary beam scanning over the sample surface.

The general ion-optical magnification of the microscope depends on the product of magnifications produced by the emission objective, transfer optics, and image converter. By varying the optical power of the transfer optics it is possible to change the locality of the analysis while keeping the size of the dissector aperture constant. In addition, it is possible to form in the aperture plane the real image of the sample surface, the image of virtual and real crossover of the emission lens. In the latter case the secondary ion beam passes completely through the aperture, and at the mass filter exit a spectrum is obtained corresponding to the sample surface composition averaged over the field of view. When the corresponding scanning voltages are applied to the deflecting plates (4), and the mass filter is set to a certain mass, the signal from the ion detector may be used (after amplification) to modulate the CRT brightness synchronously rastered with the ion image. As a result an image is obtained displaying the distribution of ions of a certain mass over the sample surface.

Development of mass spectral microscopy is far from being completed. The design of instruments already constructed has been improved to increase their optical and mass spectral resolution. The operation modes are computer controlled; spectra and images are subjected to computer processing [102]. On the other hand, the search is going on for simpler systems of ion image mass filtration, i.e. for a simpler and cheaper microscope. Here some prospects are connected with the possibility of using the focusing properties of monopole high-frequency mass filters [103].

2.7. Ion microprobes

Figure 2.14 shows the scheme of the first ion microprobe, developed by Liebl [11]. A duoplasmatron with a cold cathode serves as the primary ion source, which can generate both positive and negative ions. Primary ions are mass-separated in a 90° sector non-homogeneous magnetic field with a wedge gap between the pole pieces. The advantage of such an analyser is its stigmatic focusing and the absence of second order angular aberrations at a rather simple geometrical configuration. Within the magnet the ions follow trochoidal trajectories and leave the magnet as a parallel beam, which is collimated by the field aperture. After being separated, the beam is demagnified by two condensers through two stages. Secondary ions are collected by simple collecting optics and are transferred into a specially designed double-focusing mass spectrometer. In order to increase its transmission and make the geometrical dimensions smaller an einzel lens, which focuses the secondary ion beam, is introduced prior to the spectrometer. The secondary ion beam is energy analysed in a 45° spherical condenser followed by a 90° magnet with a wedge-type field similar to that used for the analysis of primary ions. The ion detector includes an ion–electron converter, scintillator, and PEM, and provides an equally effective registration of positive and negative ions. Two pairs of deflecting plates between the magnet and the exit field may be used for fast scanning of two neighbouring peaks in the mass spectrum, this being very useful in the measurement of isotopic ratios.

On the basis of this prototype the microanalyser IMMA was built by ARL.

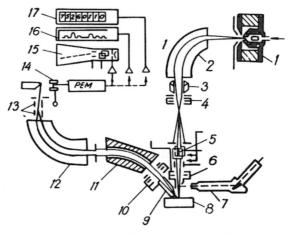

Figure 2.14 Ion microprobe mass analyzer. (1) Duoplasmatron; (2) magnetic sector; (3) deflecting system; (4) condenser; (5) deflecting system; (6) objective; (7) optical microscope; (8) sample; (9) secondary ion collection system; (10) einzel lens; (11) energy analyser; (12) magnetic sector; (13) lens; (14) ion detector; (15) CRT; (16) recorder; (17) scaler.

Subsequently, Liebl designed the combined ion and electron microprobe (UMPA) [12, 104]. In this microprobe a rather original mass spectrometer is used for secondary ion analysis. Double focusing is achieved through a 45° spherical condenser in combination with an einzel lens (which enlarges the solid angle of the ion collection) and with a 180° sector magnet (which has a wedge-type gap) for stigmatic focusing. Such a system increases mass spectral resolution $M/\Delta M$ to 250, while the energy bandpass is raised to 50 eV and the acceptance solid angle to 1.5×10^{-2} sr, which corresponds to a cone with an apex half-angle of 4°.

The instrument built by Hitachi [77] is a simplified version of the Liebl ion microprobe (Fig. 2.15). Here the primary beam is not filtered. Secondary ions are collected by a simple system and analysed in the double-focusing mass spectrometer. An interesting idea was later realized in this instrument, namely the correction of the secondary ion current for a certain mass number, taking into account the total current of secondary ions [106].

As was shown in the preceding chapter, secondary ion emission depends on some crystallographic factors and the geometry of interaction between the primary beam and the target. As a consequence, the ion image in a microscope or microprobe contains information about not only the composition, but also the structure of the target. This is especially evident in the study of fractured samples, where orientation effects from separate facets of fracture may be stronger than the effects of variations in the concentration of the element whose distribution is being studied. In order to make the compositional information free from structural superposition, it is necessary to measure at each point the total ion current and then refer to it the ion current corresponding to ion species of interest. This may be done by use of a total secondary ion current monitor (see [106]) placed between the electrostatic and magnetic sectors. In this monitor a portion of the secondary ions is

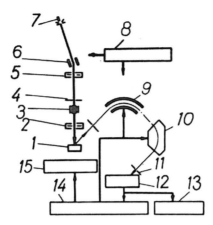

Figure 2.15 Diagram of IMA-3 microprobe. (1) Sample; (2) objective lens; (3) deflecting system; (4) objective aperture; (5) condensor lens; (6) deflector; (7) ion source; (8) pumping system; (9) electric sector; (10) magnetic sector; (11) selecting slit; (12) detector; (13) display; (14) data processor; (15) sample positioning.

captured by the aluminium plate of the ion–electron converter. Electrons ejected from the plate are accelerated and excite the scintillator, the light from which is amplified by the PEM. The signal from the PEM is fed to an analog dividing circuit, where the second input is controlled by the signal from the mass spectrometer output. As a result, the circuit measures the ratio of a filtered current to unfiltered one, and its output signal is used to modulate the CRT brightness (the CRT creates the concentration image).

The ion microprobes described above have been designed as automatic systems with specially developed mass spectrometers. However the ion microprobe can be combined with a commercial double-focusing mass spectrometer. This has been done in the AEI ion microprobe, based on a design proposed by Long *et al.* [16, 107]. Here the system for primary beam formation (it includes a cold cathode duoplasmatron, a magnetic sector, and two lenses) is made as an optional attachment to a commercial spark-source mass spectrometer, the MS–7.

The wide application of quadrupole mass spectrometers has recently inspired the development of an ion microprobe using a quadrupole base. Figure 2.16 is a diagram of a microprobe of this type. For more efficient secondary ion collection and higher transmission of a mass spectrometer channel a set of quadrupole mass analyzers may be used, arranged in a circle. The secondary ion beam optics matching the analyzer inputs may be realized using a half-cylindrical mirror [108]. The principal feature in the construction of the ion microprobe is the facility to form ion beams with high density and small cross-section [109–112]. The main limiting parameters here are the ion gun brightness and the spherical aberration of the objective lens. This is why it is much more difficult to built an ion probe than to build an electron probe, since both the parameters mentioned above are unfavourable in the case of ions. Ion guns have much lower brightness than electron guns, and the electrostatic lenses which are used for ion focusing have much stronger

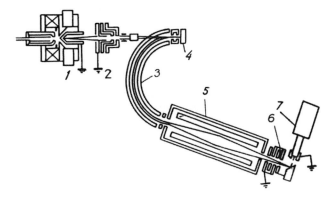

Figure 2.16 ARL ion microprobe. (1) Duoplasmatron, (2) primary beam optics; (3) spherical energy analyser; (4) sample; (5) quadrupole; (6) output optics; (7) ion detector.

spherical aberration than the magnetic lenses used for electron focusing. It should be said that in the case of duoplasmatron the spherical aberration is the factor that restricts the decrease of the probe diameter to about 2 μm. Further decrease of the beam diameter is prohibited by chromatic aberration. It must be noted, by the way, that the beam chromatic broadening is practically independent of ion energy, since the energy increase requires the proportional growth of the lens size, and this would lead to a respective rise in the chromatic aberration coefficient C_C. In principle, at a duoplasmatron brightness of 100 A cm^{-2} sr^{-1}, it is possible to get a spot 200 nm in diameter and a current of 10^{-12}A, which corresponds to a primary ion current density of about 3 mA cm^{-2}. Sources with surface ionization (they generate, for example, Cs$^+$ ions) have a much narrower ion energy spread. Hence the chromatic aberration of lenses is shown to be much weaker, and spherical aberration dominates at spot sizes up to 20 nm. For the same source brightness the current in a spot of such dimensions will amount to 10^{-13}A, which corresponds to a density of about 30 mA cm^{-2}.

In order to achieve maximum current with minimum microprobe diameter, the correct adjustment of the aperture angle is required. Figure 2.17 (thin lines) shows the dependence of the beam diameter on the ratio of beam current to source brightness for two types of sources at the optimum angle θ. But in practice this angle is fixed, and therefore diameter decrease with increased brightness growth is retarded (thick lines). Optimum operation points are those where thin lines touch thick ones. Displacement along thick lines in their upper part where the slope is about 45° corresponds to a constant current density in the spot. A slope of less than 45° implies a decrease in the current density and diminishing spot size. Thus it may be concluded, in particular, that the negative effect of a space charge, essential at large currents and with relatively broad beams, is reduced to insignificant levels when the beam diameter becomes smaller. This may be illustrated by an example: in a beam of Cs$^+$ ions at an energy of 10 keV and a current of 10^{-11}A, the average spacing between ions along the beam axis is 200 nm, but

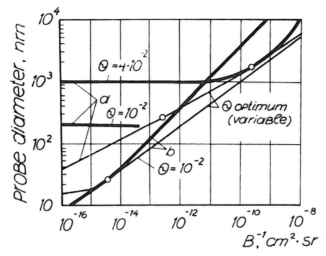

Figure 2.17 Dependence of the beam diameter on B^{-1} for the fixed aperture angle (solid lines) and the optimum angle (thin lines). (a) $- \Delta V/V = 5 \times 10^{-4}$ (duoplasmatron). (b) $-\Delta V/V = 2 \times 10^{-5}$ (surface ionization source); $C_s = 2$ cm; $C_c = 4$ cm.

this is too little for an appreciable electrostatic interaction effect. Curves of the type shown in Fig. 2.17 are useful for describing systems where the diameter of the focused beam is small compared to the ion source size. But the situation is altered when sources with a field ionization or field desorption (liquid metal ion sources) are used [113]. Such sources are extremely bright — up to six orders of magnitude brighter than a duoplasmatron. For such levels of brightness, other optics should be applied [112].

The physical limit of lateral resolution in ion probing is defined, first of all, by bombarding ion scatter in the sample (the effect is similar to that of electron scattering in electron microanalyzers). In the sample plane this scattering limit corresponds to the mean diameter of the area in which collisions initiated by an impinging ion generate secondary ions. Under conditions typical for ion microanalysers this mean diameter is about 10 nm, which is characteristic for the ultimate spatial resolution of the analysis. It must be noted that the 'informative' volume of the substance amounts in this case to $\sim 10^{-19} \mathrm{cm}^3$, this being perhaps the smallest analysable volume for any analytical method known at present. Spatial resolution is, however, limited by another important factor — the necessity to ensure a sufficient sensitivity. In other words, it is necessary to collect a definite amount of secondary ions at the mass spectrometer output to get a signal informative from the analytical point of view. Hence, while working to reduce the size of the probe, the resulting sensitivity must be kept in mind. If the concentration sensitivity is predetermined, the minimum amount of sample that must be sputtered to get a required number of registered ions may be estimated. The relationships among the parameters (relative impurity concentration C, ionization efficiency α and transmission factor K_t) for secondary ion collection system, may be conveniently described by a general sensitivity coefficient K_s : $K_s = C \alpha K_t$.

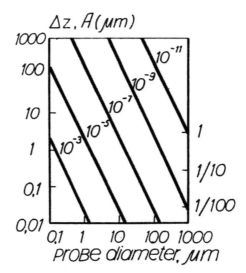

Figure 2.18 Effect of emission parameters, collection efficiency, and transmission on the thickness of the layer that must be sputtered to yield 10^2 ions (Å scale) or 10^6 ions (the same scale in μm).

Let, for example, $K_t = 10^{-1}$, $\alpha = 10^{-2}$, and $C = 1$. Then $K_s = 10^{-3}$. For $C = 10^{-2}$, $K_t = 10^{-3}$, and $\alpha = 10^{-3}$ we have $K_s = 10^{-8}$, and so on. Since the bombarded area is proportional to the square of the probe diameter, it is necessary (when the same volume of sputtered substance is to be maintained at a smaller probe diameter) to have a thicker layer sputtered during the analysis, Δz, i.e. one must sputter a layer 100 times thicker when a probe diameter 10 times smaller is required. Figure 2.18 shows the relation between the probe diameter (i.e. the lateral resolution) and the sputtered layer thickness, Δz, for various K_s (K_s values given on individual curves) necessary to get 10^2 and 10^6 ions at the mass spectrometer output. From the analysis of data shown in the figure it follows that it is extremely important to ensure the maximum transmission of the channel, where secondary ions are collected and mass analyzed, in order to obtain the optimum combination of sensitivity and lateral resolution.

References

1. Gabovich, M. D. (1964) *Ion plasma sources*. Naukova dumka, Kiev (in Russian).
2. Gabovich, M. D. (1972) *Physics and technology of ion plasma sources*. Atomizdat, Moscow (in Russian).
3. Septier, A. (1967) Production of ion beams of high intensity. In: *Focusing of charged particles*, Vol. 2, pp. 123–159, Academic Press, New York.
4. Lejeune, C. (1974) Theoretical and experimental study of the duoplasmatron ion source. *Nucl. Instrum. Methods* **116**, 417–428, 429–443.
5. Hurley, R. E. (1975) The production of focused ion beams for surface and gas-phase collision studies. *Vacuum* **25**, 143–149.
6. Sidenius, G. (1977) Gas and vapour ion sources for low-energy accelerators. In: *Low-energy ion beams*. Conference series No. 38, pp. 1–11, Bristol, London.
7. Slodzian, G. (1964) Etude d'une methode d'analyse l'emission ionique secondaire. *Ann. Phys.* **9**, 1–58.

8. Vasil'ev, M. A., Ivaschenko, Yu.N., and Cherepin, V. T. (1970) Investigation of secondary ion emission from solids by means of the mass spectrometer MI–1305. *PTE* 181–183 (in Russian).

9. Alpat'ev, Yu.S., Dubinsky, I. N., Ol'khovsky, V. L. *et al.* (1972) Mass spectrometer for the analysis of solids. *PTE* 159–160 (in Russian).

10. Liebl, H. J. and Herzog, R. F. K. (1963) Sputtering ion source for solids. *J. Appl. Phys.* **34**, 2893–2896.

11. Liebl, H. (1967) Ion microprobe mass analyzer. *J. Appl. Phys.* **38**, 5277–5283.

12. Liebl, H. (1971) Design of a combined ion and electron microprobe apparatus. *Int. J. Mass Spectrom. Ion Phys.* **6**, 401–412.

13. Liebl, H. (1975) Ion probe microanalysis. *J. Phys.* **8**, 797.

14. Liebl, H. (1974) Ion microprobe analyzers. *Anal. Chem.* **46**, 22A. The ion microprobe — instrumentation and techniques. NBS, 1975, Special Publication No. 427, SIMS, pp. 1–31.

15. Beske, H. E. (1962) Ein doppelfokussierender Massenspektrograph nach Mattauch–Herzog zur Spurenanalyse von Festkörpern. *Z. Angew. Phys.* **14**, 30–35.

16. Banner, A. E. and Stimpson, B. P. (1974) A combined ion probe spark source analysis system. *Vacuum* **24**, 511–517.

17. Barrington, A. E., Herzog, R. F. K., and Poschenrieder, W. P. (1966) Vacuum system of ion microprobe mass spectrometer. *J. Vacuum Sci. Technol.* **3**, 239–251.

18. Düsterhöft, H., Manns, R., and Rogaschewski, S. (1977) Eine Anordnung zur Messungen der positiven Sekundärionemission aus Festkörperoberflächen. *Exp. Tech. Phys.*, 117–122.

19. Jian-Hua, W., You-Zin, Li, and Pi-Jin, C. (1985) A study on argon ion gun in SIMS. In: *Secondary ion mass spectrometry, SIMS IV*, pp. 103–132, Springer, Berlin.

20. Cherepin, V. T., Dubinsky, I. N., and Dyad'kin Ya.Ya. (1982) Simple double-channel SIMS instrument. In: *Secondary ion mass spectrometry, SIMS III*, pp. 49–51, Springer, Berlin.

21. Le Goux, J. J. and Migeon, H. N. (1982) Principles and applications of a dual primary ion source and mass filter for an ion microanalyzer. In: *Secondary ion mass spectrometry, SIMS III*, pp. 52–56, Springer, Berlin.

22. Komatsu, K., Tsukakoshi, O., Katagawa, I., and Komiya S. (1984) Optimization of high brightness Cs ion source and ion optics for UHV-IMMA. In: *Secondary ion mass spectrometry, SIMS IV*, pp. 124–126, Springer Berlin.

23. Okutani, I., Fukuda, M., Noda, I., Tamura, H., and Watanabe, H. (1984) A new type surface ionization source with an additional mode of electrohydrodynamic ionization for SIMS. In: *Secondary ion mass spectrometry, SIMS IV*, pp. 127–129, Springer, Berlin.

24. Blackwell, R. J., Kubby, J. A., Lewis, G. N., and Siegel, B. M. (1985) Experimental focused ion beam system using a gaseous field ion source. *J. Vacuum Sci. Technol.* B3(1), 82–86

25. Swanson, L. W. (1983) Liquid metal ion sources: mechanism and applications. In: *Proceedings of the International Ion Engineering Congress*, ISIAT 83 – IPAT 83, Kyoto, pp. 325–336.

26. Rüdenauer F. G. *et al.* (1982) First results on a scanning ion microprobe equipped with an EHD-type indium primary ion source. In: *Secondary ion mass spectrometry, SIMS III*, pp. 43–48, Springer, Berlin.

27. Higatsberger M. J. *et al.* (1982) Operational data of a simple microfocus gun using an EHD-type indium ion source. In: *Secondary ion mass spectrometry, SIMS III*, pp. 38–42, Springer, Berlin.

28. Waugh, A. R., Bayly, A. R., and Andersen, K. (1984) SIMS with very high spatial resolution using liquid metal ion sources. In: *Secondary ion mass spectrometry, SIMS IV*, pp. 138–140, Springer, Berlin.

29. Rüdenauer, F. G. (1984) Liquid metal ion sources for scanning SIMS. In: *Secondary ion mass spectrometry, SIMS IV*, pp. 133–137, Springer, Berlin.

30. Anazawa, N. and Aihara, R. (1984) Submicron ion probes. In: *Secondary ion mass spectrometry, SIMS IV*, pp. 119–123, Springer, Berlin.

31. Liebl, H. (1984) Ion gun systems for submicron SIMS. In: *Secondary ion mass spectrometry, SIMS IV*, pp. 114–118, Springer, Berlin.

32. Franks, J. (1984) Atom beam source. *Vacuum* **34**, 259–261.

33. McDowell, R. A. and Morris, H. R. (1983) Fast atom bombardment mass spectrometry: biological analysis using an ion gun. *Int. J. Mass Spectrom. Ion Phys.* **46**, 443–446.

34. Ardenne, M. (1956) *Tabellen der Elektronenphysik, Ionen physik und Übermikroskopie*, Vol. 1, Berlin.

35. Popov, S. I. (1961) Investigation of the Ardenne duoplasmatron. *PTE*, 20–24 (in Russian).
36. Collins, L. E. and Brooker, R. L. (1962) Ion beam control in a duoplasmatron source. *Nucl. Instrum. Methods* **15**, 193–196.
37. Aberth, W. and Peterson, J. R. (1967) Characteristics of a low energy duoplasmatron negative ion source. *Rev. Sci. Instrum.* **38**, 715–750.
38. Watanabe, M. and Suita, G. (1967) Pressure-dependent characteristics of an Orkidge-type duoplasmatron ion source. *Jap. J. Appl. Phys.*, **6**, 758–764.
39. Chopra, K. L. and Randlett, M. R. (1967) Duoplasmatron ion beam source for vacuum sputtering of thin films. *Rev. Sci. Instrum.* **38**, 1147–1151.
40. Liebl, H. and Harrison W. W. (1976) Study of an iodine discharge in a duoplasmatron. *Int. J. Mass Spectrom. Ion Phys.* **22**, 237–246.
41. Aubert, J., Lejeune, C., and Tremelat, P. (1978) Beam emittance of the duoplasmatron as a function of the discharge modes. International Physics Conference Series Chapter 6, No. 38, pp. 282–286.
42. Pivovarov, A. L., Chenakin, S. P., Cherepin, V. T., and Zaporozhets, I. A. (1984). Duoplasmatron with Wien filter for surface studies. *Prib. Tekh. Eksp.* pp. 151–154 (in Russian).
43. Cherepin, V. T. (1981) *The ion probe.* Naukova dumka, Kiev (in Russian).
44. Liebl, H. (1983) Ion optics for ion microprobe instruments. *Vacuum* **33**, 525–531.
45. Ioaniviciu, D. (1973) Ion optics of a Wien filter with inhomogeneous fields. *Int. J. Mass Spectrom. Ion Phys.* **11**, 169–184.
46. Matsuda, H. (1961) Electrostatic analyser with variable focal length. *Rev. Sci. Instrum.* **32**, 850–852.
47. Liebl, H. (1984) High-resolution scanning ion microscopy and secondary-ion spectrometry: problems and solutions. Scanning Electron Microscopy Seminar, pp. 519–528.
48. Drummond, I. W. (1984) The ion optics of low-energy ion beams. *Vacuum* **34**, 51–61.
49. Slodzian, G. and Figueras, A. (1978) Remarques sur les aberrations des lentilles electrostatiques unipotentielles acceleratrices. *J. Phys. Lett.* **39**, L90–L93.
50. Shimizu, K. and Kawakatsu, H. (1974) Design charts of electrostatic three aperture lenses. *J. Phys.* **E7**, 472–476.
51. Saito, T. and Sovers, O. J. (1977) Numerical calculations of focal and aberration properties of unipotential electron lenses. *J. Appl. Phys.* **48**, 2306–2311.
52. Kanaya, K. and Baba, N. (1978) General theory of 3-electrode lenses based on the axial potential model. *J. Phys.* **E11**, 265–275.
53. Di Chio, D., Natali, S. V., and Kuyatt, C. E. (1974) Focal properties of the two-tube electrostatic lens for large and near-unity potential ratios. *Rev. Sci. Instrum.* **45**, 559–665.
54. Wittmaack, K. (1976) Successful operation of a scanning ion microscope with quadrupole mass filter. *Rev. Sci. Instrum.* **47**, 157–159.
55. Herzog, R. F. K., Poschenrieder, W. P., Liebl, H. J., and Barrington, A. (1965) Solids mass spectrometer. In: NASA Contract NASW–839: GCA Technical Report No. N67–7–n.
56. Möllenstedt, G., Benohr, A. C., and Seiler, H. (1966) Ein einfacher Massen – spektrograph zur Untersuchung der an Oberflächen ausgelösten Ionen *Z. Angew. Phys.* **20**, 5300–532.
57. Hernandez, R., Lanusse, P., Slodzian, G., and Vidal, G. (1972) Spectrographie de la masse avec source a emission ionique secondaire. *Rech. aerosp.* 313–324.
58. Hernandez, R., Vidal, G., and Lanusse, P. (1973) Analyse chimique des surfaces par spectrographie de masse a emission ionique secondaire. *Vide* 58–60.
59. Zwangobani, E. and MacDonald, R. J. (1973) Modulated, programmable mass spectrometer for ion bombardment studies. *J. Phys.* **E6**, 925–929.
60. Dowsett, M. G., King, R. M., and Parker, E. H. (1975) Modification of existing apparatus for SIMS in UHV. *J. Phys.* **E8**, 704–708.
61. Higatsberger, M. J. and Klaus, N. (1975) Ein Beitrag zum Verständnis und zur Anwendung der sekundärionen Erzeugung an Oberflächen. *Acta Phys. Austr.* **41**, 269–279.
62. Vasil'ev M. A., Krasjuk, A. D., and Cherepin, V. T. (1977) Mass spectrometer MI-1305 equipped with the ion probe and energy analyzer for investigation of solids. *PTE* 173–175. (in Russian).
63. Slobodenjuk, G. I. (1974) *Quadrupole mass spectrometers.* Atomizdat, Moscow (in Russian).
64. Sysoev, A. A. and Chupakhin, M. S. (1977) *Introduction to mass spectrometry.* Atomizdat, Moscow (in Russian).
65. Dawson, P. H. (1976) Principles of operation. In: *Quadrupole mass spectrometry and its application*, pp. 9–63, Elsevier, Amsterdam.
66. Liebl, H. (1977) Quadrupole mass filters in SIMS. *Adv. Mass Spectrom.* **7**, 418–424.

67. Krohn, V. E. (1962) Emission of negative ions from metal surfaces bombarded by positive caesium ions. *J. Appl. Phys.* **33**, 3523.
68. Benninghoven, A. and Leobach, E. (1972) Analysis of monomolecular layers of solids by the static method of secondary ion mass spectroscopy (SIMS). *J. Radioanal. Chem.* **12**, 95–100.
69. Wittmaack, K., Maul, J., and Schultz, F. (1974) A versatile in-depth analyzer. R. Backish (Ed.), *Proceedings of the 6th International Conference on Electron and Ion Beam Science and Technology*, pp. 164–171, The Electrochemical Society, Princeton.
70. Schubert, R. and Tracy, J. C. (1973) A simple inexpensive SIMS apparatus. *Rev. Sci. Instrum.* **44**, 487–491.
71. Dawson, P. H. (1977) Quadrupole for secondary ion mass spectrometry. *Int. J. Mass Spectrom. Ion Phys.* **17**, 447–467.
72. Maul, J. and Flückiger, U. (1978) Secondary ion mass spectrometry (SIMS) – a new method for the analysis of solids. *Kerntechnik*, **20**, 467–470.
73. Dawson, P. H. and Redhead, P. A. (1977) High performance SIMS system. *Rev. Sci. Instrum.* **48**, 159–167.
74. Liang-Zhen, C., Qing, X., Zhao Jia, Z., Yu Qing, T., and Young Qing, W. (1984). A retarding–accelerating energy analyzer for SIMS. In: *Secondary ion mass spectroscopy, SIMS IV*, pp. 79–81, Springer, Berlin.
75. Lehrle, R. S., Robb, J. C., and Thomas, D. W. (1962) A modified time-of-flight mass spectrometer for studying ion–molecule or neutral particle–molecule interaction. *Rev. Sci. Instrum.* **39**, 458–463.
76. Poschenrieder, W. P. (1972) Mulitple-focusing time-of-flight mass spectrometers. Part II. TOFMS with equal energy acceleration. *Int. J. Mass Spectrom. Ion Phys.* **9**, 357–73.
77. Steffens, P., Niehuis, E., Friese, T., Greifendorf, D., and Bennighoven, A. (1984) A new time-of-flight instrument for SIMS and its application to organic compounds. In: *Secondary ion mass spectrometry SIMS IV*, pp. 404–408, Springer, Berlin.
78. Vasil'ev, M. A., Chenakin, S. P., and Cherepin, V. T. (1974) The use of ion detector with the energy filter in the mass spectrometer MI–1305. *PTE*, 224–225 (in Russian).
79. Lester, J. E. (1970) Off-axis channeltron multiplier for quadrupole mass spectrometers. *Rev. Sci. Instrum.* **41**, 1513–1514.
80. Pottie, R. F., Cocke, D. L., and Gingerich, K. A. (1973) Discrimination in electron multipliers for atomic ions. II. Comparison of yields for 61 atoms. *Int. J. Mass. Spectrom. Ion Phys.* **11**, 41–48.
81. Beuhlar, R. J. and Friedman, L. (1977) Low noise, high voltage secondary emission ion detector for polyatomic ions. *Int. J. Mass. Spectrum. Ion. Phys.* **23**, 81–97.
82. Daly, N. R. (1963) New type of positive ion detector for the simultaneous measurements of two beams. *Rev. Sci. Instrum.* **34**, 1116–1120.
83. Andersen, C. A. and Hinthorne, J. R. (1972) Ion microprobe mass analyzer. *Science* **175**, 853–860.
84. Hinz, A. and Rogaschewski, S. (1977) Ein Ionen-Elektronenkonverter in Szintillationanzähler. *Exp. Tech. Phys.* **25**, 353–359.
85. Hofer, W. O. and Littmark, V. (1976) Ion and electron trajectories in mirror type ion–electron converters. *Nucl. Instrum. Methods* **138**, 67–75.
86. Hofer, W. O. and Thum, F. A. (1977) A simple axially symmetric quadrupole SIMS spectrometer. In: *Proceedings of the Ion Beam Analysis Conference*, Washington, pp. 147–163.
87. Castaing, R., Jouffrey, B., and Slodzian, G. (1960) Sur les possibilities d'analyse locale d'un echantillon par utilization de son emission ionique secondaire. *C. R. Acad. Sci, Paris* **251**, 1010–1012.
88. Cherepin, V. T. and Ol'khovsky, V. I. (1977). Ion emission microanalyzator US Patent No. 815 443. 13. VII.
89. Cherepin, V. T. and Mayfet, Yu. P. Local analysis of element concentration distribution in solids by the methods of mass spectral microscopy. Preprint 71.8, Inst. Metallofiziki AN UkrSSR, Kiev, (in Russian).
90. Cherepin, V. T. (1972) Mass spectral microscopy — a new method for investigation of solids, *Visn. Akad. Nauk. Ukr. SSR.* 17–28.
91. Cherepin, V. T. and Mayfet, Ju. P. (1972) Ion mass spectral microscopy. *Metallofizika*, **40**, 109–114 (in Russian).
92. Cherepin, V. T., Mayfet, Yu. P., and Pilipenko, A. P. Ion mass spectral microscope. *Zavod. Lab.* **39**, 484–487 (in Russian).
93. Cherepin, V. T. and Mayfet, Yu. P. (1973) Diffusion study by the method of mass spectral

microscopy. Preprint 73.3, Kiev Inst. metallofiziki AN UkrSSR, 15 pp. (in Russian).
94. Cherepin, V. T. and Mayfet, Yu. P. (1972) Ion–electron image converter. *ZhTF*, **42**, (5), 969–971 (in Russian).
95. Castaing, R., Hennequin, J.-F., Henry, L., and Slodzian, G. (1967) The magnetic prism as an optical system. In: *Focusing of charged particles*, Vol. 2, pp. 265–278, Academic Press, New York.
96. Castaing, R. (1969) Quelques applications du filtrage magnetique des vitesses en microscopie electronique. *Z. Angew. Phys.*, **27**, 171–178.
97. Borovsky, J. B., Vodovatov, F. F., Zhukov, A. A., and Cherepin, V. T. (1973) *Local methods of material analysis*. Metallurgija, Moscow (in Russian).
98. Morrison, G. H. and Slodzian, G. (1975) The ion microscope opens new vistas in many fields of science by its ability to provide spatially resolved mass analysis of solid surfaces. *Anal. Chem.* **47**, 933A–943A.
99. Slodzian, G. (1975) Looking at the collection efficiency problem through the ion microscope optics. NBS Special Publication No. 427, *Secondary Ion Mass Spectrometry*, pp. 33–61.
100. Cherepin, V. T. and Ol'khovsky, V. L. (1982) Performance and use of dissector ion microanalyzer. In: *Secondary ion mass spectrometry, SIMS III*, pp. 77–80, Springer, Berlin.
101. Möllenstedt, G. and Hubig, W. (1958) Rückstrahlung–Bildwandler für Korpuskularstableu. *Optik* **15**, 225.
102. Fassett, J. D. and Morrison, G. H. (1978) Digital image processing in ion microscope analysis: study of crystal structure effects in secondary ion mass spectrometry. *Anal. Chem.* **50**, 1861–1866.
103. Carrico, J. P. (1975) On the possibility of a mass-filter ion-emission microscope. *J. Phys.* E**8**, 18–20.
104. Liebl, H. (1978) A combined ion and electron microprobe. *Adv. Mass Spectrom.* **5**, 433–435.
105. Tamura, H., Kondo, T., and Doi, H. (1978) Analysis of thin films by ion microprobe mass analyzer. *Adv. Mass. Spectrom.* **5**, 357–363.
106. Kobayashi, H., Suzuki, K., Yukawa, K. *et al.* (1979) Correction of secondary ion intensity by a new total ion monitoring method. *Rev. Sci. Instrum.* **48**, 1298–1302.
107. Coles, J. N. and Long, J. V. P. (1978) An ion-microprobe study of the self-diffusion of Li^+ of lithium fluoride. *Phil. Mag.* **29**, 457–472.
108. Liebl, H. (1974) Quadrupole secondary ion mass spectrometry apparatus with enhanced transmission. *Int. J. Mass Spectrom. Ion Phys.* **15**, 116–119.
109. Liebl, H. (1977) Limits of lateral resolution in ion probe microanalysis. *Adv. Mass Spectrom.* **7**, 412–417.
110. Liebl, H. (1972) A coaxial combined electrostatic objective and anode lens for microprobe mass analyzer. *Vacuum* **22**, 619–621.
111. Long, J. V. P. (1965) A theoretical assessment of the possibility of selected-area mass-spectrometric analysis using a focused ion beam. *Br. J. Appl. Phys.* **16**, 1277–1284.
112. Liebl, H. (1977) Ion optics for surface analysis. In: *Low-energy ion beams*. Conference Series No. 38, pp. 266–281, Bristol, London.
113. Clampitt, Ch. and Jeffries, D. K. Molten metal field ion sources. In: *Low energy ion beams*. Conference Series No. 38 pp. 12–17, Bristol, London.

Chapter 3
Constitutional analysis of solids by SIMS

3.1. Analytical characteristics of SIMS

Mass-spectrometric investigation of sputtering products at ion bombardment is a direct way to determine elemental and isotopic composition of any solid.

SIMS like any other methods of analysis, has definite analytical characteristics. The analytic ability of a given method, that is its capability to establish an unambiguous relationship between the concentration of element under study and the collected output signal, may be described by the ion yield γ^+ (the magnitude of secondary ion current I_i^+ per unit primary current and unit concentration C_i).

If the dependence of I_i^+ is linear and unambiguous, γ^+ is described by a horizontal straight line in the $\gamma^+ - C_i$ coordinates. This simple convenience for analysis case is rather common, e.g. for metal alloys.

Differences in ionization coefficients for different elements as well as the variation of these coefficients in the case when elements form alloys, do not form an inseparable obstacle for the development of analytical methods, since calibration curves may always be plotted beforehand using alloys of a known composition as references [1].

It is a simple enough matter to find reference substances, provided complete account is taken of all the factors which influence SIE. Transition from the reference calibration curve to the measurement of an unknown concentration of the ith component is accomplished through the formula

$$C_i = K \frac{I_i^+}{I_{i(r)}^+} C_{i(r)} \tag{3.1}$$

where I_i^+ is the secondary current of the ith component in the sample; $I_{i(r)}^+$ is the secondary current of the ith component of the reference; $C_{i(r)}$ is the concentration of the ith component in the reference; and K is a factor allowing for the difference in conditions under which the comparative measurements of the sample and the reference are made.

It has been experimentally established that the secondary ion current depends linearly on the concentration of a relevant admixture in many cases. This, in particular, is valid for impurity atoms in pure materials and for dilute and concentrated solid solutions where the phase composition is not changed with the component concentration. But if the dependence $I_i^+(C_i)$ is linear the problem of references becomes much simpler, since it is sufficient to measure the currents of relevant ions in a single reference sample in order to obtain the calibration graph. Concentration dependence of the current becomes more complex when alloys with alternating phase composition are analysed. However, the non-linear character of the calibration curve at its good reproducibility does not impede the analysis in the region of macroconcentrations. However, it should be stressed that in this case SIMS is recommended for the physicochemical analysis [2–4].

Quantitative constitutional measurements by SIMS are rendered difficult by a large variation in SIE coefficients for different elements. Also, the absolute values of these coefficients depend strongly on the experimental conditions, this being the cause of the wide spread in the data available from the literature.

To circumvent these difficulties it is necessary, using outer and inner references, to determine relative sensitivity factors (RSF) for a number of elements and systems.

In the case of SIMS this factor may be expressed through the ratio of the secondary ion yield for a definite element to that for the reference.

When secondary ion currents are measured by mass spectral means it is convenient to determine the RSF not for all ions of the ith element but for its most abundant isotope. Then the ion yield of the ith element may be given as

$$\gamma_{i(n)}^+ = I_{i(n)}^+/I_0 C_i K_t \tag{3.2}$$

Here $I_{i(n)}^+$ is the peak current of nth isotope of ith element measured at the mass spectrometer collector; I_0 is the primary beam current; C_i is the ith element concentration; and K_t is the mass spectrometer transmission. Hence,

$$\text{RSF} = \frac{\gamma_{i(n)}^+}{\gamma_{r(n)}^+} = \frac{I_{i(n)}^+ I_0 K_{t(r)} C_{i(r)}}{I_{r(n)}^+ I_0 C_i K_t} \tag{3.3}$$

Here r implies reference. At constant values of I_0 and when $K_{t(r)} = K_t$ and $C_i = C_{i(r)}$,

$$\text{RSF} = I_{i(n)}^+/I_{r(n)}^+ \tag{3.4}$$

It is not difficult to find from Equation (3.4) the minimum value of a measured concentration δ, that is the sensitivity limit of the analysis:

$$\delta = I_{min}^+/I_r^+ \, \text{RSF} \tag{3.5}$$

where I_{min}^+ is the minimum detected secondary ion current. We have determined RSF and δ values in the mass spectrometer MI-1305 equipped with an ion probe. Pure metals were studied at Ar^+ and He^+ bombardment (8 keV, 1.8 mA cm^{-2}) [5, 6]. RSF and δ values for the most abundant isotopes of 45 pure metals bombarded by Ar^+, He^+, and O_2^+ ions are summarized in

Table 3.1. Elements are shown in order of decreasing RSF for Ar^+ bombardment. Iron bombarded by Ar^+ ions serves as a reference. As may be seen from the table, RSF values may differ by three orders of magnitude in the case of Ar^+ bombardment, three and a half orders for He^+ bombardment and four orders for O_2^+ bombardment. This is quite natural and determines the sensitivity range of the analysis: from 1.9×10^{-7} to $5 \times 10^{-3}\%$. This

Table 3.1
Relative sensitivity factors (RSF) and detection limit δ for various elements

Element	RSF			$\delta, 10^{-5}\%$		
	Ar^+	He^+	O_2^+	Ar^+	He^+	O_2^+
Lu	15.6	2.64	16.6	0.40	3.3	0.527
Mg	11.4	13.3	86	0.55	0.65	0.101
Al	9.62	62.2	447	0.65	0.14	0.019
Sc	7.75	31	205	0.81	0.28	0.042
Tb	7.00	3.63	23.8	0.89	2.4	0.367
Er	4.50	2.03	15.8	1.39	4,27	0.554
Nd	4.13	1.11	17.8	1.51	7.8	0.49
Ho	2.88	3.03	21.4	2.18	2.9	0.41
Y	2.87	8.4	51.2	2.19	1.04	0.17
V	2.75	13.1	200	2.28	0.66	0.044
Cd	2.28	1.47	13.2	2.75	5.9	0.66
Hf	2.25	1.33	8.58	2.78	6.5	1.02
Be	2.19	15.5	163	2.86	0.65	0.053
Mn	2.15	0.95	13.5	2.90	9.1	0.645
Dy	1.88	2.84	23.1	3.33	3.06	0.38
Tm	1.68	6.7	22.1	3.73	1.3	0.396
Ti	1.56	7.14	127	4.0	1.21	0.069
In	1.50	30	53.8	4.17	0.29	0.162
Co	1.22	0.37	5.97	5.1	23.4	1.46
Nb	1.16	4.05	35.6	5.38	2.15	0.245
Yb	1.05	2.8	25.2	5.92	3.1	0.35
Pr	1.03	2.61	22.2	6.1	3.33	0.396
Cr	1.01	2.95	41.6	6.25	2.96	0.21
Fe	1.00	0.872	15.8	6.25	10	0.555
Ni	0.98	0.415	4.42	6.41	21	1.98
Rh	0.89	0.556	24	7.05	15.4	0.364
Ru	0.76	0.724	20.6	8.26	12.1	0.426
Re	0.65	2.14	20.5	9.6	4.08	0.351
La	0.49	0.725	12.8	12.8	12	0.68
Ce	0.40	0.773	14.7	15.6	11.2	0.59
Mo	0.385	1.86	25.2	16.3	4.67	0.35
Bi	0.358	0.13	0.464	17.5	67	18.9
Zr	0.296	2.08	28.2	21.2	4.2	0.31
Sm	0.275	1.84	19.5	22.8	4.73	0.45
Cu	0.262	0.235	1.26	23.8	37	6.95
W	0.187	0.477	5.83	33.3	18.3	1.5
Ta	0.186	0.853	5.43	33.4	10.2	1.61
Pt	0.112	0.013	0.076	56	667	114
Ag	0.108	0.07	0.135	58	123	64.6
Sn	0.094	0.202	0.558	66.7	43	15.7
Pb	0.086	1.08	1.22	72.5	8.05	7.15
Pd	0.042	0.033	1.55	147	262	5.63
Zn	0.034	0.107	0.918	185	81	9.5
Cd	0.019	0.016	0.106	333	547	82.7
Au	0.012	0.025	0.051	500	350	172

sensitivity is probably not a limit. In [7] a number of factors have been theoretically considered which determine the sensitivity limit of mass spectrometers where the ion sputtering is used. It is shown that in principle a mass spectrometer possessing a secondary ion current up to 10^{-8} A is possible. This corresponds to a dynamic range of nine orders of magnitude when secondary ion detectors with a low noise level are used. Under ideal conditions (the absence of mutual superposition of lines in spectrum, adverse effect of residual gases, of the primary beam and of the sputtering of electrodes, large differences in atomic masses under study, etc.) it may be possible to measure concentrations of less than 10^{-8}% (10^{-1} p.p.b.) depending on the type of element.

It must be mentioned that this high sensitivity is attainable first of all in the analysis of such elements as potassium, sodium, calcium, and lithium, this being of extreme importance for analysis of raw materials and products in semiconductor technology.

Negative secondary ion emission, which is as yet poorly understood, is important for analytical aims. Yields of negative secondary ions are higher for non-metals than for metals [11]. This fact may be successfully used in the analysis of such impurities as carbon, oxygen, chlorine, and the like. Under normal conditions the emission of negative ions from metals is not high, but it may be greatly enhanced (by an order of magnitude) through bombardment with caesium ions or inert gas ions when the metal surface is covered by a continuously recovering caesium film [9, 10, 12, 13].

Application of negative SIMS with Cs^+ ion bombardment makes is possible to increase the number of elements reliably measured with low detection limits in solids, this being most effective in the case of As, Si, Sb, Se, Te, Au, and Pt [14]. Negative SIMS of O, C and S is preferable with inert gas ion bombardment, since Cs^+ ion bombardment results in a high background level. The detection limit for some impurities may be lowered if secondary cluster ions are measured. The type of ion is defined by a particular combination of matrix and impurity element. While measuring, for example, Ge and O in GaAs, detection of $^{149}AsGe^-$ ions instead of $^{74}Ge^-$, and $^{85}GaO^-$ instead of $^{16}O^-$, makes it possible to lower detection limits for these elements by 65 and 10 times, respectively.

Surface and bulk analysis of solids by means of SIMS may become troublesome or impossible if insulating samples are to be examined. The problems arise from charge build-up on the specimen, which causes not only a shift in the total energy of the emitted particles (with respect to some reference potential) but also a change in the trajectories of both primary and secondary ions. Charging of the specimen prevents meaningful results, since mass spectrometry always involves some kind of energy analysis. The mass scale in magnetic-sector-type instruments may be shifted, and resolution of the quadrupoles may degrade as the energy of the secondaries rises.

The most effective way to compensate for primary ion charge is to use an auxiliary electron beam. It has been demonstrated that not only positive but also negative secondary ions from insulators can be recorded with the help of charge-compensating electrons. It is desirable to use well-defined, 0.5–2.0 keV, electron beams. It is interesting to note that charge compensation was better achieved under conditions when the electron current was much larger than the ion current, whereas the opposite was true for respective current densities [15].

In view of the large excess of electrons required for charge compensation, the appropriate term for this technique is electron-*beam* flooding or spraying, as opposed to flooding with *thermal* electrons. In the latter case, one allows for partial charging of an insulating sample, which in turn generates the potential necessary for attracting electrons from a heated filament. The technique may be not adequate for SIMS analysis, in particular if the insulator contains dopants of interest which become mobile as a result of the residual surface charge.

The mechanism of charge compensation by electron beam flooding is likely to depend upon the electron energy and may involve some production of electron–hole pairs, which in turn could promote the flow of current. Several other methods have been also suggested and tested for various insulating samples: surface coating withu vapour-deposited conducting films, deposited or impregnated grids, negative oxygen or negative iodine ion bombardment, sample bias voltage adjustment, and placement of a conducting diaphragm on the surface [16–18].

3.2. Qualitative constitutional analysis

Raw data for an analysis in SIMS generally are acquired in the form of a mass scan, i.e. a record of the spectrometer detector current as a function of scanning time. Three steps of data reduction have to be performed before elemental concentration values can be obtained:

(1) Peak search. This step locates the peaks in the sampled mass scan and generates a list of peak heights (areas) vs mass number in low-resolution mass scans. This list, called the 'mass spectrum', may also include peaks at non-integer mass numbers originating from doubly or triply charged atomic or molecular ions.

(2) Peak interpretation. In this step, the peaks of a mass spectrum are interpreted in terms of contributions from monoatomic ions, isoelemental clusters, and heteroatomic molecular ions, taking into account isotopic and molecular overlap at integer mass numbers. As a result of this step, a list of identified ionic species and corresponding ion currents, summed over all possible isotopic combinations of each species, is produced.

It is appropriate to mention here that many of the problems connected with overlapping mass number can be avoided when using a high-resolution mass spectrometer. The advantage of unambiguous peak interpretation however can be offset by increased instrument cost and analysis time, and by reduced analytical sensitivity.

(3) Elemental quantification; this final step generates a list of identified elements and corresponding fractional or absolute atomic concentrations (atom cm^{-3}) from the list of identified ionic species.

The first two steps are the objective of qualitative analysis and are considered below in more detail.

The main problems in the interpretation of low-resolution SIMS spectra originate from the fact that they are not simple linear superpositions of spectra from separate elements, because of large differences in RSF and matrix effects. Different energy distributions for ions of different types as well as structural effects are another source of trouble.

On the other hand, the fact that the natural isotopic abundance remains more or less stable in a secondary ion mass spectrum [19], is very favourable and is the basis for reliable identification of corresponding peaks.

Peak interpretation is most frequently performed 'manually' on the basis of general information on the formation of simple and complex ions and information contained in the Periodic Table.

'Manual' interpretation is frequently susceptible to human error, quantitative inaccuracy, and subjective judgement, especially on the composition of polyatomic ions.

A simple method for obtaining information on combinations of elements whose ions may correspond to particular spectral lines is suggested in [20]. However selection among these combinations must be done by the operator and at least an approximate composition and nature of the sample must be known for this purpose.

In an analytical laboratory it is desirable to automate this whole procedure, so that the operator has to interfere only when absolutely necessary. At present several computer programs have been proposed for the interpretation of mass spectra of completely or partially unknown samples [21–26].

The first algorithm of this type was developed by Steiger and Rüdenauer [21] and was implemented in the computer code SIP (Spectrum Identification Program). The program begins with the search for an element whose most abundant isotope corresponds to the most prominent peak in the mass spectrum. Then the maximyum ion currents that can be assigned to isotope ions of this element are determined. The highest possible values of isotopic currents are subtracted (stripped) at the proper mass numbers from the original spectrum, yielding a 'residual' spectrum. In this residual spectrum the elemental stripping is repeated iteratively. Following the identification of a predetermined number of elements, SIP similarly looks for clusters, oxides, and hydroxides of these elements, stripping again the maximum possible isotopic currents from the residual spectrum.

In the next step binary, ternary, and quaternary hybrid molecules of previously identified elements may be stripped. This procedure is repeated until a predetermined number of peaks have been classified as 'unidentifiable'.

This method contains a great deal of uncertainty, since the final result depends on the choice of initial conditions and control parameters. However, the results are remarkably good when the computer solution is compared to 'manual' interpretation data.

There is another possibility for formal interpretation of mass spectra from samples of unknown composition [22, 23]. When the general elemental composition of a sample is known, a spectrum interpretation must be started with identification of lines at mass numbers corresponding to the most abundant isotopes of its main components. Interpretation for polyisotopic elements is checked by comparing the height of the corresponding peaks with the natural isotope abundance of these elements. At the second stage the multiply charged (usually doubly charged) ions of the main components are identified. Their mass number in the spectrum (the *m/e* ratio) is half that of the relevant singly-charged ions. The currents of doubly charged ions are one to three orders of magnitude lower than those of singly charged ions (except in special cases).

The identification of isoelemental clusters starts with the search for two-atomic combinations of main component atoms, allowing for equal probability for combinations of different isotopes of a given element. For example, in case of the Ti_2^+ ion, the natural abundance of the mass number 96 is calculated as the sum of the products of natural abundances of isotope pairs ^{46}Ti, ^{50}Ti, ^{47}Ti, and ^{48}Ti. The number of peaks in the mass spectrum is accordingly increased and amounts to five for Ti^+, nine for Ti_2^+, thirteen for Ti_3^+, etc.

For bombardment regimes most common in SIMS (Ar^+ or O_2^+ ions with a current density of 1–2 mA cm^{-2} at the pressure of residual gases 10^{-4}–10^{-5} Pa) the emission intensity of polyatomic ions is usually decreased with the growth of the number of atoms in the cluster.

Further interpretation includes identification of peaks which correspond to typical SIMS molecular and cluster ions — e.g., M_xO_y oxides, M_xH_y hydrides and M_xO_yH hydroxides — as well as to hybrid molecules M_xN_y. Production of complex ions containing oxygen and hydrogen atoms depends significantly on the nature of the target and on experimental conditions of analysis.

A simple method for obtaining information on combinations of chemical elements whose ions may correspond to a particular spectral line is suggested

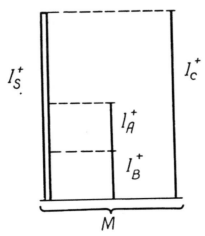

Figure 3.1 Diagram of the mass spectrum decoding.

in [16]. However, a choice between these combinations must be made by the operator, and at least the approximate composition and nature of the sample must be known for this purpose. In order to determine the origin of a given line in the spectrum one must have at hand preliminary collected information on combinations and superpositions of chemical elements whose ions can correspond to a particular spectral line.

Application of computers is highly promising for the development of techniques of qualitative analysis. Thus in [17] a computer program has been proposed for alternative identification of lines in mass spectra.

Types of ions which are more preferable for description of a mass spectrum may be found from the calculation of the criteria D. The meaning of these criteria may be understood from Fig. 3.1. Let some spectral line S be partly or completely identified from contributions of ions of isotopes A, B, and C possessing the same mass number. As may be seen from Fig. 3.1, if the sum of the highest possible ion currents for isotopes A and B ($I_A^+ + I_B^+$) is less than the intensity of the spectral line I_S^+, the C isotope corresponds actually to the S line because otherwise this cannot be identified without a residual part remaining.

In practice it is more convenient to calculate not the sum but the ratio of the ion current sum to the spectral line intensity. Such a criterion for e.g., C ions, equals $D_c = (I_A^+ + I_B^+)/I_S^+ = 0.5$ (Fig. 3.1). In a similar way the criteria are calculated for A and B ions: $D_A = D_B = 1.25$. The magnitude of the criterion D determines to what extent a given ion is more or less preferable for the description of the spectral line than the whole assembly of other ions suitable for identification of this line. The lower the value of D, the more appropriate for the mass spectrum is a given ion.

It should be noted that mathematical interpretation of mass spectra is equivalent to the solving of an indefinite system of linear equations where the left-hand parts are represented by spectral line intensities, while the right-hand parts are given by the sums of unknown ion currents for isotopes possessing the same mass numbers. For example in the case shown in Fig. 3.1, this system is reduced to a single equation with three unknown ion currents: $I_S^+ = X_A^+ + X_B^+ + X_C^+$, where $0 \leqslant X_{A,B,C}^+$. The use of a simple x method in the mathematical programming makes it possible to choose a solution of the underdetermined system that satisfies the minimum of some function relating quantities of interest. The function being minimized is composed in such a way that includes unknown ion currents X^+ along with D criteria, calculated beforehand, and unidentified residual spectral intensities. A solution must be found which corresponds to the minimum number of D criteria and to minimum residuals. This implies that the results of the spectrum interpretation contain information about the most preferable ions (minimization of criteria) for the most complete (without any residual) interpretation of the mass spectrum (minimization of residuals).

The matrix method, developed by Antal *et al.* [24, 25] and implemented in the computer code MATRIX, is an attempt to put the identification procedure on a mathematically sound basis (for details see original papers). In

MATRIX there are also a number of free control parameters which, to some extent, allow the operator to influence the results of the interpretation. The two identification algorithms, MATRIX and SIP, have been tested by applying them to the same experimentally recorded mass spectra, and have yielded comparable results in the identification of SIMS and liquid–metal ion source mass spectra [25].

One more so-called probability method has been proposed recently [26]. Basically it is similar to the repetitive spectrum stripping method but during each cycle the most prominent peak of an elemental ion is looked for along with all corresponding cluster ions, in particular, oxides, hydrides, and hydroxides. This block is subtracted and the procedure is repeated until the background noise level is reached. The authors claim, that the approach saves much time and may be realized in a microcomputer, taking only 2–4 min for interpretation of a crowded mass spectrum.

A fundamental difficulty, inherent to low-resolution mass spectrum interpretation, is the necessity of relying upon natural isotopic abundances for the identification of an element or molecule. Measured isotopic abundances may deviate from natural abundances for various reasons: there may be isotopic fractionization in the sample previous to analysis; in the course of the analysis there may be isotopic effects in sputtering and ionization; and instrumental mass discrimination effects may occur, particularly in quadrupole mass spectrometers. A decisive increase in identification accuracy may be expected when more physical information on the process of ion formation, analysis, and detection is fed into the interpretation algorithm.

3.3. Physical background of quantitative analysis

Application of SIMS as a method for quantitative analysis is based on the knowledge of SIE regularities and the mechanism of ion formation. Unfortunately, quantum-mechanical models which treat the ion emission as a combination of processes taking place in sub-surface layers and close to the surface do not lead to quantitative results. These models can provide qualitative interpretation of the emission from rather simple objects, such as metals, binary alloys, and simple compounds [27–36]. But quantitative interpretation of the raw data from more complex materials is difficult because there is insufficient consistency in the realization of most models, and there is a lack of the data required for physical characterisation of the surface. Therefore, in the analytical sense the SIE thermodynamical models are more appropriate. They are based on the assumption of the existence in the sputtering region of equilibrium and non-equilibrium local plasmas [37–39], composed of neutrals and also of ionized atoms, electrons, and photons. These models are not concerned with problems of particle motion through sub-surface layers in the target, and the emission characteristics are not related to the nature of the sample material. It is supposed that the effects which can appear due to the action of the target crystal lattice are completely suppressed, while the ion emission is enhanced and stabilized by the presence of chemically active gases introduced into the area where ions are formed.

From the point of view of the quantitative analysis, the SIE thermodynamic model proposed by Andersen and Hinthorne [38] is regarded as the most promising. According to this model the particles to be analysed are in a state of thermodynamic equilibrium with the surface. This state resembles a dense plasma and is characterized by the plasma temperature T and the electron density N_e. The equilibrium state results mainly from a high density of particles in the sputtering region: this density is supposed to be 10^{20}–10^{21}cm^{-3} for electrons. Processes where atomic and molecular ions are produced may be represented by the relevant dissociation reactions. For instance, positively charged monoatomic ion can be produced as a result of a neutral atom dissociation, by the reaction $M^0 \rightleftarrows M^+ + e$ with the decomposition constant

$$K_M^+ = N_M^+ N_e / N_M^0 \qquad (3.6)$$

where N_M^+, N_M^0 and N_e are concentrations of ions, neutrals, and electrons per unit volume, respectively.

The partitition constant may be found from the Saha–Eggert equation, which together with Equation (3.6) allows calculation of the ionization degree of the element M atoms:

$$\frac{N_M^+}{N_M^0} = \frac{A T^{\frac{1}{2}} B_M^+}{N_e B_M^0} \exp\left[-(I_M - \Delta E)/kT\right] \qquad (3.7)$$

where A is a constant, B_M^+ and B_M^0 are the internal partition functions of the ionized and neutral atoms, respectively; I_M is the ionization potential of the sputtered atom; and ΔE is the ionization potential depression due to Coulomb interaction of the charged particles.

Similar equations may be deduced for the formation of negatively charged ions and molecular ions of MO and MO$_2$ type.

Equation (3.7) includes two unknown quantities: the temperature T and density of electrons N_e. These values may be found from the solution of Equation (3.7) for two elements contained in the sample in known concentrations. Obtained in this way T and N_e are then used in Equation (3.7) to calculate the concentrations of other elements. Practical application of the Andersen and Hinthorne model gives quite good results but it has since been shown that physical parameters of SIE do not correspond to the notion about collective particle interaction in the plasma. Secondary ion energy distribution differs essentially from the Maxwell distribution, and the most probable ion energies (several eV) correspond to temperatures that are incompatible with T values calculated from the Saha–Eggert equation [39]. These facts suggest that the processes which occur in the region of the surface being sputtered do not agree with the assumption of existence of a thermal equilibrium, and that Equation (3.7) is only an empirical formula with fitting parameters T and N_e.

The model proposed in [30] assumes that the electron gas is in a state of equilibrium with the surface, which has temperature T and electron work function $e\varphi$. The relation between ionized and neutral atoms is given by the Sacha–Eggert equation:

$$\frac{N_M^+}{N_M^+} = \frac{B_M^+}{B_M^0} \exp \left[-\frac{(I_M - \varphi)}{kT} \right] \qquad (3.8)$$

However, the applicability of this model for SIE description remains questionable. Substitution of real T and φ values into Equation (3.8) gives N_M^+/N_M^0 values that exceed by many orders of magnitude the measured value. In order to explain experimental data the temperature must be raised to a level above the critical temperature, that is T and φ must be taken as the fitting parameters. Urela believed that the time interval when the equilibrium state is established may be longer than the time interval when ionization of the atom occurs. Therefore it is reasonable to consider ion emission as thermodynamically unsteady process where the degree of atom ionization follows Dobretsov's equation:

$$\frac{N_M^+}{N_M^0} = \frac{B_M^+}{B_M^0} \exp \left[-\frac{(I_M - \_\_E - .)}{kT} \right] \qquad (3.9)$$

It is easy to see that Equations (3.7)–(3.9) are qualitatively similar. They become completely identical under the condition

$$kT\ln (AT^{3/2}/N^e) + \Delta E = \varphi \text{ or } \varphi + \Delta E \qquad (3.10)$$

where the terms are taken from Equations (3.7)–(3.9).

In other words, if the characteristics shown in Equation (3.10) are treated as the fitting parameters, the quantitative analyses performed on the basis of each approximation must lead to identical results. This is supported by experimental data.

3.4. Methods of quantitative analysis

The expression for the concentration of element M for known currents of its singly charged ions has the form

$$N_M = \frac{N_M^+}{N_M^+/N_M^0} \left[\left(1 + \frac{N_M^+ + N_M^-}{N_M^0} \right) \left(1 + \frac{N_O^0}{K_{N_1}^0} + \frac{(N_O^0)^2}{K_{N_1}^0 K_{N_2}^0} \right) \right] \qquad (3.11)$$

here N_M^+ and N_M^- refer to experimentally measured currents of positive and negative ions of the element M and also to the currents of molecular ions of the MO^+ and MO_2^+ type. Concentrations of neutral atoms of element M and oxygen, as well as the partition constants $K_{N_1}^0$ and $K_{N_2}^0$ may be given by relevant equations as a function of T and N_e. Calculations with Equation (3.11) are performed using the CARISMA computer program [38]. T and N_e values are obtained from solution of two Equations (3.11) deduced for two reference samples with known concentrations. In practice it is necessary to find a matrix of values for T and N_e parameters giving correct concentration values for the reference elements. These parameters are then used in (3.11) to calculate concentration of other elements in the sample. The sum of ion currents of all the elements is normalized to 100%. The concentration corresponding to each element is calculated as the quantity proportional to the appropriate normalized ion current.

Methods for decreasing the number of reference elements are also proposed in [38]. A long analytical practice resulted in the discovery of an empirical relationship between T and N_e values ($\log T = 2.817 + 0.0638 \log N_e$). As a consequence, the number of unknowns has been reduced to one, and only one inner reference element can be used. A similar relation ($N_e = -44.15 + 15.67 \log T$) was suggested in [40]. It has been also found that T and N_e parameters do not change essentially if the samples under study are similar in physical nature and chemical composition. Hence the reference-free analysis of typical samples is possible at fixed T and N_e values.

Further improvement of these methods proceeds in the direction of introduction and approval of assumptions for the simplification of the calculation procedure. For example, if the contribution of all ions except those of M^+ is ignored, then Equation (3.11) is reduced to Equation (3.7). The results obtained by means of Equation (3.7) with a simplified program of CARISMA type are discussed in [40]. It has been shown that if ΔE in Equation (3.7) is variable or even equal to zero, this does not affect the accuracy of analysis. A similar effect is produced by approximating the distribution function by a polynomial of the fifth degree or by temperature-independent parameters. These assumptions make it possible to significantly decrease the number of parameters from the periodic system that are used to represent information about elements in the computer memory. As a result the program is simplified, and the analysis is made accordingly faster.

Variation of parameters entering Equation (3.7) in the algorithm of the QUASIE program [41, 42] led to the conclusion that the accuracy of analysis is not reduced essentially when the function is approproximated by a third-order polynomial or by parameters independent of temperature and electron density. The contribution from ΔE in Equation (3.7) may be ignored. It has been also shown that the distribution functions may be substituted by their constant values taken at 5000 K or approximated at around 4500 K. The latter fact may perhaps be responsible for the results obtained with the use of Equation (3.9) in [43], where the distribution functions for the fixed temperature 5100 K were applied, this being of no large consequence for the accuracy of the standard sample analysis.

SIMS quantitative analysis may be performed on experimental data without involving any mechanism of ion formation. Various modifications of such an approach (with the use of reference or calibration straight lines [44], inner references, calculation of ion relative intensities, and sensitivity factors [46–50] are based on the same idea: that the experimentally measured current of atomic ions is proportional under certain conditions to the concentration of a relevant element in the sample.

A vast amount of practical material and experience in quantitative SIMS of various objects of natural and artificial origin has been accumulated to date. Many works are devoted to the analysis of insulators, such as glass, minerals, and other geological objects of terrestrial and lunar origin [35, 38, 44], analysis of semiconductor compounds [38, 44, 51], carbides, borides, phosphids, sulphides, silicates [4, 38, 39, 41, 55], metals and metallic alloys [1–3, 42, 44, 46, 56–60], and in particular stainless and low-alloyed steels [35, 41, 43, 45, 47, 61, 62]. Techniques have been developed to identify hydrogen [63, 64], rare earth elements [65], iodine [66], and to analyse the composition of dust microparticles in the air [67], and in aluminium alloys [68]. The methods of isotope analysis have also been developed [19, 69, 70].

SIMS has been successfully applied to the study of environmental samples. The results of these studies are of considerable importance from the standpoint of environmental chemistry and demonstrate the need for quantitative determination both of particle surface concentrations and of elemental depth profiles. Thus, for example, depth profile analyses of electric steel furnace dust particles were performed and surface concentrations were calculated with matrix-corrected sensitivity coefficients. Bulk concentrations of the dust particles were calculated using SIMS and tube-excited energy dispersive X-ray fluorescence analysis. The results have shown considerable enrichment of the dust particle surface by such elements as Cr, Mn, Co, Cu, Zn, and Pb [74], which are responsible for the chemical activity and toxicity of these particles. Quantitative analysis was most successful with the use of electronegative gases (mainly oxygen and chlorine) either as primary ions or as the active background in the residual gas atmosphere. For illustration, Fig. 3.2 shows results obtained in [38], where the content of transition metals and boron in matrices of different metals and silicates was determined. Oxygen ions served as the primary species. More than one hundred different standards were analysed in a wide range of concentrations, from 10^{-4} to tens of atomic per cent. The ions $^{16}O^-$, $^{18}O^-$, and $^{35}Cl^-$ were tried as bombarding species. The straight lines in the figure are the lines of ideal correlation with

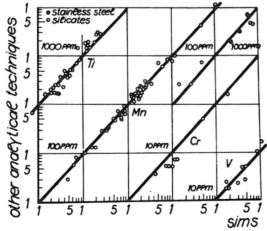

Figure 3.2 Comparison of the analysis of the ion microprobe mass analyser with those of other analytical techniques for some of transition metals in silicate matrices plus B in silicates and metals. After [38].

the data from other standard methods of quantitative analysis. The relative error in the method does not exceed 10% for most elements under consideration. Equation (3.11) was used in calculations within the frameworks of the CARISMA program.

The emission process stimulated by the action of chemically active elements is known in the literature as chemical or reaction emission, and it is a powerful tool in quantitative analysis, since increasing secondary ion current corresponds to a rise in SIMS sensitivity (for elements with high oxygen affinity SIMS sensitivity can reach 10^{-5}–10^{-8} atomic%). On the other hand, stabilization or saturation of secondary ion currents at a definite pressure of active gases helps to increase the accuracy and reproducibility of results [64]. Figure 3.3 shows the data obtained from the analysis of a stainless steel sample bombarded with Ar^+ [Fig. 3.3(a)] and O_2^+ (Fig. 3.3(b)) ions in a vacuum of 10^{-5}Pa [54]. It becomes evident from the comparison of these figures that the quantitative analysis may be successfully achieved only with the bombardment by oxygen ions. However, as has been stressed in [61], the best results were obtained at complete saturation of secondary ion currents due to additional inlet of oxygen. Elemental concentrations were calculated in [54] within the framework of a two-reference simplified method with the use of Equation (3.7).

The effect of oxygen on the results of the analysis is most pronounced at oxygen pressures of 10^{-6}–10^{-3} Pa. At lower pressures the reaction emission does not essentially effect the magnitude of the secondary ion currents, even if inert gas ions are used for bombardment [41]. But in the latter case one must take into account possible effects connected with the physico-chemical nature of the surface being bombarded. The crystal lattice effect, known also as the 'matrix effect', may be so great that it cannot be suppressed completely with the use of active gas only as bombarding ions. Thus Fig. 3.4 shows calibration curves obtained from the analysis of iron base alloys bombarded

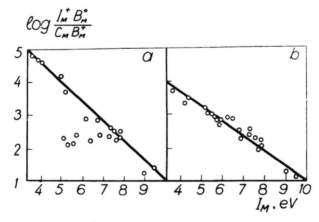

Figure 3.3. Dependence of $I_M^+ B_M^0 / C_M B_M^+$ on the ionization potential I_M of various elements in NBS 461 stainless steel matrix for Ar^+ (a) and O_2^+ (b) ion bombardment in vacuum 10^{-5}Pa. (a) $T = 7880 \pm 260$ K. (b) $T = 10\,900 \pm 690$ K.

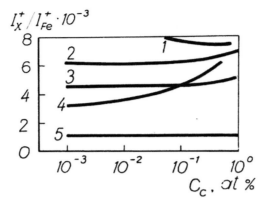

Figure 3.4 Effect of carbide formation on the results of analysis of alloying elements in steel. (1) Fe–Cr–C; (2) Fe–V–C; (3) Fe–Nb–C; (4) Fe–Mn–C; (5) Fe–Ni–C.

with O_2^+ ions in a vacuum of 10^{-5} Pa [45]. The data in this figure are in the form of the dependence of relative intensities of component ions on carbon concentration. It may be seen that the experimental conditions of the analysis do not provide complete suppression of effects connected with carbide formation in high-carbon-content steels. In Fig. 3.4 this effect appears as a departure from linearity in the concentration dependence of ion relative intensities. This abnormality may be used as the basis for physico-chemical analysis [1–3].

The use of oxygen primary ions with simultaneous oxygen flooding at pressures above 10^{-3} Pa make it possible to successfully analyse similar steels [40] using the method of reference curves. The alloying element concentration can be varied from 10^{-3} to tens of atomic%.

In the majority of works where the effect of the reaction emission has been taken into account it has been pointed out that relative error in SIMS analysis can amount to 10–40%, depending on the element under investigation. This accuracy is quite comparable with those of other standard methods of quantitative analysis such as AES, electron probe, wet chemical analysis etc. It has been established that the energy variation in bombarding ions does not lead to a marked alteration in the results. At the same time, the accuracy of the analysis may be increased by shifting the energy interval of the detected secondary ions. Thus Fig. 3.5(a) represents measurements in the energy window 0–20 eV. Weak data correlation is evident. Energy window shifting to 40–60 eV makes accuracy of the analysis much higher due to the smaller effect of molecular ions, whose energy distribution is significantly narrower than that of atomic ions [Fig. 3.5(b)]. (See also [72].)

Control of reaction emission processes greatly improves the dynamic range of concentration measurement. SIMS sensitivity to different elements depends on the chemical nature of the elements and on the nature of the matrix material. These factors determine the ultimate values of minimum detection limits, which may differ by several orders of magnitude. As an example, Table 3.2 summarizes detection limits for Na, Al, Si, and Fe contained in

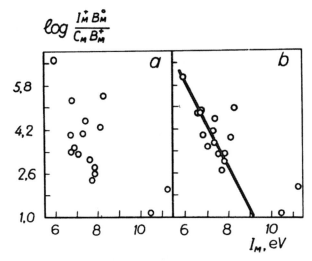

Figure 3.5 Effect of the energy 'window' position for secondary ions on the accuracy of quantitative analysis. (a) Window 0–20 eV. (b) Window 40–60 eV. PbF_2 matrix.

Table 3.2
Detection limits of elements ($C_{min} \times 10^{-4}$ weight%) at different oxygen pressures [73]

Admixture	Matrix							
	Nb	Ti	Cu	Zn	Nb	Ti	Cu	Zn
	$P_{O_2} < 2 \times 10^{-5}$ Pa				$P_{O_2} = 1 \times 10^{-3}$ Pa			
Na	0.001	0.001	0.002	0.001	0.001	0.001	0.002	0.001
Al	0.2	0.4	0.2	0.5	0.002	0.003	0.03	0.5
Si	5	7	10	5	0.05	0.05	1	5
Fe	20	13	15	20	0.2	0.1	3	20

matrices of various metals. The data refer to two fixed oxygen pressures in the mass spectrometer chamber [73]. It may be seen that the detection limits may be lowered if the oxygen pressure is increased to 10^{-3} Pa. But further increase of the oxygen pressure may have adverse effects on the sensitivity of the method, due to possible recombination of ionized particles with oxygen atoms.

3.5. Analysis of organic and biological samples

The detection, identification, and structural determination of organic compounds, very often available only in extremely small total quantities, is an important analytical problem in general organic chemistry, biochemistry, medicine, environmental control, and other areas. A wide variety of organic materials are of interest in these fields e.g., amino acids, peptides, drugs, vitamins, pharmaceuticals, herbicides, and pesticides. SIMS has recently become a very important technique for the detection and analysis of this type of large involatile organic molecule. It is applied to solid and liquid samples, which are bombarded by charged or neutral atomic or molecular primary particles [75–82]. All these different modes of secondary ion mass spectrometry are based on the same, still poorly understood phenomenon: the emission of large undestroyed molecular secondary ions during sputtering of a corresponding sample [79].

There are four typical ion groups in the mass spectrum of secondary emission from organic compounds of complex composition:

(1) quasimolecular ions of the $(M - OH)^+$ and $(M - H)^+$ type for efedrine, $(M - H)^-$ for nicotine acid, $(M - H)^+$, and $(M - H)^-$ for sulphanilamide (where M is the mass number of the corresponding molecule);

(2) secondary ions of the substrate $(Ag^+, AgCl_2^-)$. Their appearance indicates incomplete coverage of the substrate with the substance under analysis;

(3) characteristic positive and negative ions of large molecular fragments;

(4) ions of small fragments of mass number below 100.

For analytical purposes, ions of the first type are most representative. Their concentration is directly proportional to the amount of the corresponding substance. Secondary ion emission coefficients turned out to be extremely high for these ions. Table 3.3 gives the values for 18 amino acids, plus some peptides, medical drugs, and vitamins.

Measurements have shown the reproducibility of spectra to be within a factor of two when samples are prepared using a micropipette. The detection limit is 10^{-14} g for a known substance and 10^{-9} g for an unknown one that requires identification of mass spectrum lines.

A very efficient ionization of the sample molecules M during sputtering is achieved if these molecules are deposited as a monolayer on a noble metal surface. The application of a noble metal as substrate material avoids the destruction of the sample molecules by surface reactions [79].

In general, the monolayer deposition of a molecular species M on a noble metal is very simple. If an appropriate solution is deposited on the surface, which has previously been cleaned or prepared by etching, ion pre-bombardment, or by evaporation, crystallite precipitation in general occurs only after the formation of a closed monolayer of the sample molecules and the substrate. In a typical sample preparation procedure for quadrupole SIMS about 1 µl of a 10^{-3} mol l^{-1} solution of the sample molecules is deposited on 0.1 cm^2 of the substrate surface, resulting in about one monolayer. If a lower concentration is applied, submonolayers are formed.

One cm^2 of such a monolayer contains some 10^{14} molecules. This is the total amount of material that can be sputtered from this monolayer. It limits the number of parent ions that can be produced during analysis. The situation does not improve essentially if the sample molecules are deposited as multilayers on the substrate, since during sputtering of the uppermost monolayer, subjacent lower layers are dramatically changed by radiation damage. As a result, a drastic decrease of molecular ion emission occurs if these deeper layers are sputtered.

A breakthrough in organic SIMS with low transmission instruments was the

Table 3.3
Absolute yields $S(X)$ of "parent-like" secondary ions of amino acids and various organic compounds on silver. Primary ions: Ar^+, 2.25 keV, primary ion dose density 4×10^{-6} A s cm^{-2} [78].

Compound	Formula	Yield $S(X) \times 100$		
		$(M + H)^+$	$(M - H)^-$	$(M - COOH)^+$
Glycine	$C_2H_5NO_2$	120.0	—	52.0
A-alanine	$C_3H_7NO_2$	21.0	40.0	53.0
B-alanine	$C_3H_7NO_2$	88.0	19.5	7.2
Phenylalanine	$C_9H_{11}NO_2$	4.0	0.3	13.0
Serine	$C_3H_7NO_3$	61.0	18.0	61.0
Threonine	$C_4H_9NO_3$	8.3	1.6	13.8
Proline	$C_5H_9NO_2$	19.2	8.8	72.0
Valine	$C_5H_{11}NO_2$	8.0	8.3	32.0
Leucine	$C_6H_{13}NO_2$	0.8	26.4	40.0
Norleucine	$C_6H_{13}NO_2$	24.8	6.5	76.0
Arginine	$C_6H_{14}N_4O_2$	7.2	2.4	2.1
Tyrosine	$C_9H_{11}NO_3$	7.4	—	13.6
Tryptophan	$C_{11}H_{12}N_2O_2$	3.5	0.8	3.5
Cysteine	$C_3H_7NO_2S$	12.0	11.0	15.0
Cystine	$C_6H_{12}N_2O_4S_2$	4.0	1.6	1.8
Methionine	$C_5H_{11}No_2S$	13.1	5.4	9.4
Ethionine	$C_6H_{13}NO_2S$	13.6	5.6	12.0
Glutamine	$C_5H_{10}N_2O_3$	7.2	8.3	4.3
Peptides:	$C_6H_{11}N_3O_4$	4.0	0.4	2.0
Glycylglycylglycine	$C_8H_{16}N_2O_3$	1.6	4.2	3.0
Glycylleucine				
Drugs:		$(M + H)^+$	$(M - N)^-$	$(N - OH)^+$
Barbital	$C_8H_{12}N_2O_3$	—	44.0	—
Ephedrine	$C_{10}H_{15}NO$	16.0	—	40.0
Epinephrine	$C_9H_{13}NO_3$	—	6.4	—
Vitamins:		$(M + H)^+$	$(M - H)$	
Ascorbic acid (C)	$C_6H_8O_6$	3.7	17.6	
Biotin (H)	$C_{10}H_{16}N_2O_3S$	0.3	4.2	
Nicotinic acid (PP)	$C_6H_5NO_2$	—	46.4	

deposition of a relatively large quantity of sample molecules in a regenerating glycerol matrix (see references in [79]). Glycerol has a very low vapour pressure, in the 10^{-3} Pa range, and hence it does not disturb the operation of the UHV pumping system. Typically a glycerol droplet of some μl loaded with some μg of the sample material is deposited on an area of some 0.01 cm^2 of a sample holder.

Due to and depending on the surface activity of the sample molecules, surface enrichment occurs in the uppermost monolayers of the glycerol matrix, very often resulting in a closed monolayer of these molecules on the surface even at a very low volume concentration. Ion bombardment results in sputtering of the sample molecules out of this segregation layer. But due to mobility of the sample molecules in the glycerol, a continuous regeneration of this surface layer occurs, and even at very high primary particle flow densities, corresponding to removal of several monolayer equivalents in one second, a relatively high and stable secondary ion emission can be obtained, provided a sufficient amount of sample material has originally been deposited in the matrix.

The results of a systematic investigation of this segregation process for a number of organic compounds show very clearly that the sputtering yields and the secondary ion yields are both very similar if the same sample molecules are sputtered from a glycerol or from a silver substrate, at least for the sample molecules investigated (cetrimonium bromide, atropine, sucrose and fructose) [79, 81].

Emission of the sample molecules can be initiated by atom or cluster bombardment of the target. The bombarding particles may be neutral or charged. The charge state has no influence on the secondary ion yield, but ion beams are preferable, since they can be easily deflected, focused, mass-separated and measured. Neutral particle bombardment may be useful in the case of insulators to alleviate surface charge-up, but this kind of primary particle causes serious problems in beam deflection and focussing, and for current measurement.

Neutral primary particles (FAB) have been used in the first SIMS experiments with double-focusing mass spectrometers, and they are still applied in the majority of current organic SIMS investigations using this kind of instruments. But the actual reason for using neutral bombarding particles is not the charging effect of the liquid target. As a matter of fact, there is no considerable charging of the glycerol matrix for the current densities applied, and neutral particles are used to overcome the difficulty of bringing a beam of low-energy charged particles to the target, which is kept at high voltage in the secondary ion source of the spectrometer. Appropriate ion optics can avoid these difficulties and make all advantages of ion beam available for SIMS with double-focusing instruments [81]. Investigations of the ion formation process suffer from the fact that samples are usually prepared from solution under atmospheric conditions. Various reactions between the substrate material and components of the solution or the atmosphere may occur, and contamination of the substrate surface cannot always be avoided. To reduce the number of possible interactions, it is therefore necessary to perform sample preparation under UHV conditions.

It has been shown [83, 84] that the preparation of amino acid samples in the mono- and multilayer range is possible under UHV conditions by using a molecular beam technique. The molecular beam source was calibrated by a quartz crystal microbalance, and layer thickness was determined by AES in the static mode. This preparation technique has been applied in a systematic investigation of secondary ion emission from amino acid overlayers on various metals. An extension of these experiments to other kinds of organic substances with relatively low vapour pressures seems very promising.

Considerable increase in sensitivity of analysis may be achieved using Hg FAB [83] or liquid metal ion sources [85]. Several LMIS were tested (Ga, In, Si–Au entectic alloy, and Bi). These sources operated stably even in the presence of relatively high vapour pressure organics (10^{-3}Pa). Quasi-molecular ions were produced from all of the test compound with very high abundances, an order of magnitude or greater over achieved using fast Ar atom bombardment. There was no evidence in the mass spectra of any compounds for liquid metal interaction with the sample. The analytical capabilities of the LMIS are further enhanced by practical features such as size, simplicity of operation in the unfocused mode, vacuum compatibility. The relative abundances of the quasi-molecular ions were observed to be approximately Ar(1) : Ga(10) : In(40) : SiAu(50) , i.e. about a 50-fold increase in secondary ion formation efficiency when changing from Ar to SiAu.

Another promising method of analysis of organic materials is the structural analysis of various polymers and other high molecular weight compounds in combination with determination of their chemical composition. Since all substances of this kind are insulators, alleviation of the sample surface charging is of principal importance in SIMS analysis. This may be accomplished either by the use of fast neutral atoms, which are produced when corresponding ions are charge-exchanged on gas molecules [86, 87], or by compensation of the surface charging by low-energy electron flooding.

It is also important to make the sample surface free from hydrocarbon contaminations which produce ions of the same type as those from the sample under study. It has been shown in [87] that when polymers of the polyethylene type are analysed, the sample surface may be cleaned of contaminations by heating to 150°C. For other polymeric materials similar heating is also sufficient, since their mass spectra remain practically unchanged with further temperature raise.

The Ar^+ ions or fast Ar atoms are used for bombardment with an energy of 1 keV and fluence of 10^9–10^{10} atom $cm^{-2}s^{-1}$. Secondary ion spectra depend weakly on the primary energy in the range 200–2000 eV.

Especially successful have been attempts aimed at microstructural analysis of fluoropolymers via determination of mutual orientation of monomeric cells in macromolecular co-polymers [86, 88, 89]. Thus for example, it has been shown in [88] that the molecule of polythreefluoroethylene has predominantly an ordered structure and is constructed in a 'head to tail' fashion. Polyfluorovinilidene has a mixed structure where the 'head to tail' structure dominates. The development of processing techniques for the determination of the relative intensities of mass spectral lines from ions of CX_3^- type from copolymers containing fluorine, made it possible to solve practically important problems of the chemistry of high molecular weight compounds [89].

SIMS of various substances that under normal conditions are in a liquid state is also of interest. They may be analysed after conversion into a solid state through freezing. The charged cluster emission was studied in this manner from the films of mixtures of different isotopic water modifications [90, 91], as well all the cluster emission from water on ions of alkali metals [92].

Investigation and analysis of organic and inorganic dielectrics, various frozen liquids, and gases has only just begun, but the first results have already found many useful analytical applications [94–96]. However, the physical mechanism by which the ions of these substances are formed is still rather unclear. Efforts in this direction should lead to new important information on the nature of the interaction between charged particles and complex molecular species.

3.6. Standards and cross-calibration of instruments

In the evolution of a basis for quantitative compositional analysis by SIMS, the development of standards plays a critical role. As has been shown in previous sections, a 'first principles' approach to SIMS analysis which relies on calculation alone is difficult to develop, and the practical analyst is thus dependent upon standards for quantitative analysis [97].

The principal characteristics of a suitable standard for determination of relative sensitivity factors (RSF) for SIMS are:

(1) The composition should be homogeneous, both laterally and in depth, since the sample volume eroded during determination of the RSF may be quite shallow. Alternatively, an ion-implanted standard can be used if the composition profile is known, or if the total dose is known and signals can be integrated over the profile to give a single measure of the RSF.

(2) The composition should not be significantly altered during ion bombardment and must be stable over long periods of time.

(3) The structure of a standard should not affect the results of RSF measurement; single-phase homogeneous and amorphous solids are most desirable.

As a result of the extraordinary flexibility of SIMS analysis, a wide variety of standards are needed to satisfy the diverse analysis problems that may be encountered. The development of such standards, however, requires considerable expenditure of effort and resources. Moreover, in developing a suitable standard, the greatest problem encountered is often the lack of an independent artifact-free reference method which can be used to calibrate the SIMS standard. Neutron activation and X-ray fluorescence analysis are often useful for the measurement of trace and minor constituents in a standard, but these techniques provide little information on the spatial distribution.

Electron-probe X-ray and Auger electron spectroscopy microanalysis have lateral and (or) depth resolution similar to SIMS, but they do not have adequate sensitivity at the trace level. Complete characterization of SIMS standards may necessitate employing multiple analysis techniques, including SIMS itself, which can be used to determine lateral and depth homogeneity while a bulk technique provides an independent measure of concentration.

It should be mentioned that even highest level standardization materials, e.g., NBS Standard Reference Material (SRM) must be used with caution for SIMS. SRMs are usually created for a particular purpose—such as bulk analysis, in which a relatively large quantity of the standard is consumed for each analysis—and are not necessarily suitable for a local analysis technique like SIMS, in which the sample consumption is orders of magnitude lower.

It has therefore proved necessary to develop special Standard Reference

Materials which can serve as SIMS standards [97]. These include: (1) bulk glasses; (2) special geometric forms; and (3) a depth profiling standard.

Glasses in which the constituents are combined as oxides have a number of advantages as SIMS standards. Glasses can be made homogeneous on the nanometer scale. The constituents are present in a single phase which is non-crystalline, thus eliminating orientation effects. The high oxygen content (0.4–0.65 mass fraction) helps to establish an oxygen saturation condition. Glasses can be readily fabricated in special shapes, including fibres, spherical and irregular particles, and thin films.

Special geometrical forms must be used as standards for the analysis of individual small particles by SIMS. These forms include a large proportion of environmental and technological particles based on oxides.

Ion-implanted standards have proven extremely useful for quantitative SIMS analysis of electronic materials [98]. The difficulties in making such standards for SIMS include accurately calibrating the total dose, and determining the depth distribution by an artifact-free independent profiling technique. Unfortunately the complementary depth-profiling techniques either depend on sputtering (Auger electron spectroscopy) or the use of high-energy beams (Rutherford backscattering) and introduce considerable radiation damage even if sputtering is avoided. The implementation of the neutron depth profiling technique has made it possible to prepare standards with profiles measured with the minimum energy deposition in the sample [97].

Another side of the standardization problem refers to the fact that sensitivity factors incorporate directly the instrumental response factors of the particular instrument on which they were measured. As such, the need for measuring or calculating the instrument response is obviated when these sensitivity factors are used to analyse an unknown. However, such sensitivity factors cannot be transferred to another instrument, and may be inappropriate for the same instrument with any modification of the operating parameters. These parameters include bombardment, and environmental and instrumental characteristics; they have a decisive influence on the value of the RSF of a particular element. It is obvious that all these parameters should be identical for any two SIMS instruments in order to guarantee identical mass spectral intensity ratios and hence identical, (and therefore 'transferable') RSFs between those instruments. Such instruments may be called 'cross calibrated' instruments [99].

The idea of cross-calibration relies upon a hypothesis that if two instruments, under identical bombardment and environmental conditions, produce identical RSFs from a calibration sample, their windows in secondary ion energy and directional distribution are identical; they therefore should also produce identical RSFs from samples other than the calibration sample.

The most important point in the cross-calibration procedure is selection of an appropriate 'primary calibration standard' (PCS). This standard should meet the requirements already indicated. In addition, it should contain at least three elements with standardized concentrations so that at least two independent peak height ratios may be measured; these elements should cover as wide a mass range as possible so that mass dependence of instrument transmission (in particular of quadrupole spectrometers) can be efficiently cancelled out. The secondary ion energy distribution of the elements contained should be sufficiently different so that relative sensitivity factors of at least two elements are sensitive to variations in energy bandpass of the instrument: this allows adjustment of the energy bandpass of a particular SIMS instrument by measuring RSFs.

Metallic glasses of approximate composition $B_{20}Fe_{80}$ with addition of a minor third element have been chosen for these reasons. In particular the system $B_{15}Fe_{75}W_{10}$ meets the requirements for a primary calibration standard given above.

An actual cross-calibration which has been carried out on seven different instruments in various laboratories in several countries has shown excellent results. An average agreement within a factor of $<1, 7>$ has been achieved in the determination of RSF [99].

References

1. Cherepin, V. T. (1971) Mass spectrometric measurements of secondary ion emission from alloy as an analytical tool. *Adv. Mass Spectrom.* **5**, 448–450.
2. Cherepin V. T. and Vasil'ev M. A. (1977) Secondary ion emission from concentrated alloys and compounds. In: *Proceedings of the VII International Conference on Atomic Collisions in Solids, Moscow*, p. K7.
3. Cherepin, V. T. (1978) Secondary ion mass spectrometry of metals and alloys. *Adv. Mass Spectrom.* **7**, 776–783.
4. Cherepin, V. T., Kosyachkov, A. A., and Vasil'ev, M. A. (1978) Secondary ion emission of transition metal carbides. *Phys. Status Solidi* **A50**, K113–K116.
5. Vasil'ev, M. A., Chenakin, S. P., and Cherepin, V. T. (1974) Relative sensitivity of analysis performed by the method of ion–ion emission (in Russian). *Dokl. Akad. Nauk Ukr. SSR Ser. A.*, No. 8, 751–753.
6. Vasil'ev, M. A., Chenakin, S. P., and Cherepin, V. T. (1975) Determination of relative yield coefficients for secondary ions. *Anal. Khim.* **30**, (3), 611–612 (in Russian).
7. Rüdenauer, F. G. (1971) Some basic considerations concerning the sensitivity of sputtering ion mass spectrometers. *Int. J. Mass Spectrom. Ion Phys.* **6**, 309–323.
8. Sloane, R. H. and Press, R. (1938) The formation of negative ions by positive ion impact on surfaces. *Proc. R. Soc.* **A168**, 284–301.
9. Krohn, V. E. (1962) Emission of negative ions from metal surfaces bombarded by positive caesium ions. *J. Appl. Phys.* **33**, 3523–3526.
10. Abdulaeva, M. K., Ajukhanov, A. Kh., Gafurova, M., and Shamsiev, G. V. (1972) The study of negative ion sputtering of some metals irradiated by Cs ions. in *Atomnye stolknovenija na poverkhnosti tverdogo tela*, pp. 3–15, FAN, Tashkent (in Russian).
11. Yamaguchi, N., Suzuki, K., Sato, K., and Tamura, H. (1979) Determination of carbon in steel by secondary negative ion mass spectrometry. *Anal. Chem.* **51**, 695–698.
12. Ajukhanov, A. Kh. and Turmashev, E. (1979) Determination of negative ionization degree at sputtering. *Zh. Tekh. Fiz.* **49**, (6) 1234–1237.
13. Bernheim, M. and Slodzian, G. (1977) Caesium flooding on metal surfaces and sputtering negative ion yield. *J. Phys. Lett.* **38**, L325–L328.
14. Li, A. G., Lototskiy, A. G., Gimel'farb, F. A., Orlov, P. V., Antonova, E., and Borodina, O. M. (1982) Application of negative secondary ion mass spectrometry to analysis of surface layers of solids. *Poverkhnost'*, 94–100 (in Russian).
15. Wittmaack, K. (1979) Primary-ion charge compensation in SIMS analysis of insulators. *J. Appl. Phys.* **50**, 493–497.
16. Blanchard, B., Carrier, P., Hilleret, N. *et al.* (1976) Utilization des canons a électrons pour l'étude des isolants a l'analyseur ionique. *Analysis* **4**, 180–184.
17. Andersen, C. A., Roden, H. J., and Robinson, C. F. (1969) Negative ion bombardment of insulators to alleviate surface charge up. *J. Appl. Phys.* **40**, 3419–3420.

18. Ganjei, J. D. and Morrison, G. H. (1978) Quantitative ion probe analysis of glasses by empirical calibration methods. *Anal. Chem.* **50**, 2034–2039.
19. Okano, J., Nishimura, H., and Ochiai, T. (1983) Isotope effects due to ion bombardment in SIMS. *Proc. Int. Ion Engineering Congr. ISIAT-83 & IPAT 83*, Kyoto, p. 1929–1940.
20. Gries, W. H. (1975) Slide-rule for quick identification of superimposed mass-spectral lines. *Int. J. Mass Spectrom. Ion Phys.* **18**, 272–274.
21. Steiger, W. and Rüdenauer, F. G. (1975) A computer program for peak identification in secondary ion mass spectra. *Vacuum* **25**, 409–413.
22. Cherepin, V. T., Kosyachkov, A. A., and Gudzenko, G. I. (1980) Decoding of ion mass spectra with computer aid *Zh. Anal. Khim.* **35**, 283–287 (in Russian).
23. Cherepin, V. T., Kosyachkov, A. A., and Gudzenko, G. I. (1980) Computer peak identification in SIMS. *Int. J. Mass Spctrom. Ion Phys.* **35**, 225–230.
24. Antal, J., Kuler, S., and Ridel, M. (1982) Computer peak identification and evaluation of SIMS spectra. In: *Secondary ion mass spectrometry, SIMS III*, pp. 297–300, Springer, Berlin.
25. Steiger, W., Rüdenauer, F. G., Antal, J., and Kugler, S. (1983) Automated peak interpretation in low-resolution SIMS spectra; a comparison of two algorithms. *Vacuum* **33**, 321.
26. Batalov, B. V., Kunilov, V. A., Kolomeytsev, M. I., and Naumov, A. E. (1985) Method for qualitative analysis of secondary ion mass spectra. *Poverkhnost'* (in Russian). (in press).
27. Benninghoven, A. (1969) Zum Mechanismus der Ionenbildung und Ionenemission bei der Festkörperzerstaubung. *Z. Phys.* **220**, 159–180.
28. Cherepin, V. T. (1975) Ion beam interaction with metal surface In: *Metally, elektrony, reshetka*, pp. 294–311, Naukova dumka, Kiev (in Russian).
29. Padzersky, V. A. and Zypinjuk, B. A. (1978) Atom ionization close to metal surface. *FTT* **11**, 3283–3287 (in Russian).
30. Jurela, Z. (1975) The application of non-equilibrium surface ionization to the emission of secondary ions. *Int. J. Mass Spectrom. Ion Phys.* **17**, 77–88.
31. Antal, J. (1976) On the quantum theory of the emission of secondary ions. *Phys. Lett.* **55A**, 493–494.
32. Cini, M. (1976) A new theory of SIMS at metal surfaces. *Surf. Sci.*, **54**, 71–78.
33. Düsterhoft, H., Manns, R., and Hilderbrandt, D. (1976) Experimental proof of a theoretical relation for the probability of positive ion excitation by bombardment of solid surfaces with 12 keV Ar^+ ions. *Phys. Status Solidi* **A36**, K93–K97.
34. Sroubek, Z., Zavadil, J., Kubec, F., and Zdansky, K. (1978) Model of ionization of atoms sputtered from solids. *Surf. Sci.* **77**, 603–614.
35. Smith, D. H. and Christie, W. H. (1978) A comparison of a theoretical model and sensitivity factor calculations for quantification of SIMS data. *Int. J. Mass Spectrom. Ion Phys.* **26**, 61–76.
36. Blaise, G. and Slodzian, G. (1974) Evolution des rendements de l'emission ionique des alliages avec la nature an solute premier partie: resultats experimentaux. *J. Phys.* **35**, 237–241.
37. Andersen, C. A. (1970) Analytic methods for the ion microprobe mass analyzer. Part. II. *Int. J. Mass Spectrom. Ion Phys.*, **3**, 413–428.
38. Andersen, C. A. and Hinthorne, J. R. (1973) Thermodynamic approach to the quantitative interpretation of sputtered ion mass spectra. *Anal. Chem.* **45**, 1421–1438.
39. Andersen, C. A. (1975) A critical discussion of the local thermal equilibrium model for the quantitative correction of sputtered ion intensities. NBS Special Publication No. 427, *Secondary ion mass spectrometry*, pp. 79–119.
40. Simons, D. S., Beaker, J. E., and Evans, C. A. (1976) Evaluation of the local thermal equilibrium model for quantitative secondary ion mass spectrometric analysis. *Anal. Chem.* **48**, 1341–1348.
41. Rüdenauer, F. G., Steiger, W., and Werner, H. W. (1976) On the use of the Saha–Eggert equation for quantitative SIMS analysis using Ar primary ions. *Surf. Sci.*, **54**, 553–560.
42. Rüdenauer, F. G. and Steiger, W. (1976) Quantitative evaluation of SIMS spectra using Saha–Eggert type equations. *Vacuum*, **26**, 537–543.
43. Lototsky, A. G. and Gimel'farb, F. A. (1976) Quantitative interpretation of mass spectra of secondary ion emission for the determination of composition of solids. *Zh. Anal. Khim.* **31**, (3), 433 (in Russian).
44. Tsunoyama, K., Ohashi, J., and Suzuki, T. (1976) Quantitative analysis of low alloy steels with the ion microprobe mass analyzer. *Anal. Chem.* **48**, 833–836.
45. Ishitani, T., Tamura, H., and Kondo, T. (1975) Quantitative analysis with an ion microanalyzer. *Anal. Chem.* **47**, 1294–1296.

46. Gerlach, R. L. and Davis, L. E. (1977) Semiquantitative analysis of alloys with SIMS. *J. Vacuum Sci. Technol.* **14**, 339–342.
47. Servais, I. P., Graas, H., Leroy, V., and Habraken, L. (1976) Analyse quantitative de la surface de l'acier par microanalyse ionique. *Vide,* **31**, 27–29.
48. Snowdon, K. J. (1978) A comparison of experimental secondary ion energy spectra of polycrystalline metals with theory. *Radiat. Eff.* **38**, 141–149.
49. Socha, A. J. (1971) Analysis of surface utilizing sputter ion source instruments. *Surf. Sci.* **25**, 147–170.
50. Slodzian, G. (1975) Some problems encountered in secondary ion emission applied to elementary analysis. *Surf. Sci.* **48**, 161–186.
51. Huber, A. M. and Moulin, M. (1972) Use of the ion microanalyzer for the characterization of bulk and epitaxial silicon and gallium arsenide. *J. Radioanal. Chem.* **12**, 75.
52. Vasil'ev M. A., Ivaschenko, Ju. N., Chenakin, S. P., and Cherepin, V. T. (1973) Investigation of secondary ion emission from CdTe. In: *Tr.XV Vsesojuznoy konf. po emiss.elektronike,* 19–22 nojabrja, Kiev, Ch. 2, pp. 172–174 (in Russian).
53. Vasil'ev, M. A. and Zhukov, A. G. (1974) The use of secondary ion–ion emission method for the investigation of semiconductive materials. In: *Vzaimodejstvie atomnyh chastits s tverdym telom,* pp. 210–212, Naukova dumka, Kiev (in Russian).
54. Vasil'ev, M. A., Zavalin, I. V., Maksimov, V. K. *et al.* (1974) Secondary ion–ion emission of thermodiffused impurities in thin CdP_2 layers. In: *Vzaimodejstvie atomnykh chastits s tverdym telom,* Naukova Dumka, Kiev, pp. 213–214, (in Russian).
55. Vasil'ev, M. A., Kosyachkov, A. A., Nemoshkalenko, I. N., and Cherepin, V. T. (1976) Peculiarities in secondary ion–ion emission from transition metal borides. In: *Vzaimodejstvie atomnykh chastits s tverdym telom,* pp. 106–109, Izd-vo Khar'kov, Khar'kov (in Russian).
56. Vasil'ev, M. A. and Muktepavel, F. O. (1978) Investigation of fracture surfaces of Al and Pb compounds with Sn obtained by cold welding. *Metallofizika,* **71**, 12–15 (in Russian).
57. Riedel, M., Nenadovič, T., and Perovič, B. (1978) SIMS study of iron–nickel and iron–chromium alloys. I, II, III. *Acta Chem. Acad. Sci. Hung.* **97**, 177–185, 187–196, 197–206.
58. Pickering, H. W. Ion sputtering of alloys. *J. Vac. Sci. Technol.* **13**, 618–621.
59. Pivin, J. C., Roques-Carmes, C., and Slodzian, G. (1978) Variation des rendements d'emission ionique secondaire des alliages Ni–Cr, Fe–Ni, Fe–Cr en fonction de la teneur en solute. *Int. J. Mass Spectrom. Ion Phys.* **26**, 219–235.
60. Vasil'ev, M. A., Koval', J. N., Kosyachkov, A. A., and Cherepin, V. T. (1977) Quantitative analysis of Fe–Ni alloys by the secondary ion emission method. *Dokl. Akad. Nauk Ukr. SSR* Ser. A, No. 4, 354–356 (in Russian).
61. Morgan, A. E. and Werner, H. W. (1976) Quantitative analysis of low alloy steels by secondary ion mass spectrometry. *Anal. Chem.* **48**, 699.
62. Schelten, J. (1963) Massenspektrometrische Untersuchung der Sekundärionenemission von Legierungen. *Z. Naturforsch.* **23a**, 109–113.
63. Benninghoven, A., Müller, K. H., and Schemmer M. (1978) Hydrogen detection by SIMS: hydrogen on polycrystalline vanadium. *Surf. Sci.* **78**, 565–576.
64. Pavlyak, F., Bori, L., Giber, J., and Buhl, R. (1977) Detection of hydrogen in metals by the SIMS method with quadrupole mass filter. *Jap. J. Appl. Phys.* **16**, 335–342.
65. Ishizuka, T. (1974) Secondary ion mass spectrometry of rare earth elements. *Anal. Chem.* 1487–1491.
66. McHugh, J. A. and Sheffield, J. C. (1965) Mass analysis of subnanogram quantities of iodine. *Anal. Chem.* **37**, 1099.
67. McHugh, J. A. and Stevens, J. F. (1972) Elemental analysis of single micrometer size airborne particulates by ion microprobe mass spectrometry. *Anal. Chem.* **44**, 2187–2192.
68. Poschenrieder, W. P., Herzog, R. F., and Barrington, A. E. (1965) The relative abundance of the lithium isotopes in the Holbrook meteorite. *Geochim. Cosmochim. Acta.* **29**, 1193.
69. Westerman, E. J. (1970) The ion microprobe mass spectrometer in Al alloy research. *J. Metals* **22**, 28–31.
70. White, F. A., Sheffield, J. C., and Rouke, F. M. (1962) Isotopic abundance determination of copper by sputtering. *J. Appl. Phys.* **33**, 2915.
71. Schubert, R. (1974) The effect of oxygen adsorption on the positive secondary ion yield of stainless steel. *J. Vac. Sci. Technol.* **11**, 903–905.
72. Steele, J. M., Hutcheon, J. D., and Solberg, N. T. (1977) Effect of energy selection on quantitative analysis in secondary ion microanalysis. *Int. J. Mass Spectrom. Ion Phys.* **23**, 293–305.
73. Himmelfarb, F. A., Kovarsky, A. P., Li, A. G., and Orlov, P. B. (1980) Reaction emission effect on the results of solid surface analysis by the method of secondary ion mass

spectrometry. *J. Anal. Khim.* **25**, 213–223.
74. Van Craen, M., Natusch, D. F. S., and Adams, F. (1982) Quantitative surface analysis of steel furnace dust particles by secondary ion mass spectrometry. *Anal. Chem.* **54**, 1788–1792.
75. Benninghoven, A., Jaspers, D., and Sichtermann, W. (1976) Secondary ion emission of amino acids. *Appl. Phys.* **11**, 35–39.
76. Benninghoven, A. and Sichtermann, W. (1977) Secondary ion mass spectrometry; a new analytical technique for biologically important compounds. *Org. Mass. Spectrom.* **12**, 595–597.
77. Benninghoven, A. and Sichtermann, W. (1978) Detection, identification and structural investigation of biologically important compounds by secondary ion mass spectrometry. *Anal. Chem.* **50**, 1180–1184.
78. Benninghoven, A. (1979) Organic secondary ion mass spectrometry. NBS Special Publication No. 519, *Trace organic analysis; a new frontier in analytical chemistry*, pp. 627–635.
79. Benninghoven, A. (1984). Organic secondary ion mass spectrometry. In: *Secondary ion mass spectrometry, SIMS IV*, Springer, Berlin, pp. 342–356.
80. Benninghoven, A. (1983) Some aspects of secondary ion mass spectrometry of organic compounds. *Int. J. Mass Spectrom. Ion Phys.* **53**, 85–99.
81. Benninghoven, A. Organic secondary ion mass spectrometry (SIMS) and its relation to fast atom bombardment (FAB). *Int. J. Mass Spectrom. Ion Phys.* **46**, 459–462.
82. Stoll, R., Shade, V., Rollgen, F. W., Giessmann, V., and Barofsky, D. F. (1982) Fast atom and ion bombardment of organic samples using mercury. *Int. J. Mass Spectrom. Ion Phys.* **43**, 227–229.
83. Lange, W., Holtkamp, D., Jirikowsky, M., and Benninghoven, A. Secondary ion emission from UHV-deposited amino acid overlayers on metals.
84. Holtkamp, D., Lange, W., Jirikowsky, M., and Benninghoven, A. (1984) UHV preparation of organic overlayers by a molecular beam technique. *Appl. Surf. Sci.* **17**, 296–308.
85. Barofsky, D. F., Giessmann, V., Swanson, L. W., and Bell, A. E. (1983) Molecular SIMS with a liquid metal field ion point source. *Int. J. Mass Spectrom. Ion Phys.* **46**, 495–497.
86. Tantsyrev, G. D. and Kleimenov, N. A. (1973) The use of atom–ion emission for the mass spectral analysis of F-polymers. *Dokl. Akad. Nauk SSSR* **213**, 649–652 (in Russian).
87. Tantsyrev, G. D. and Povolotskaja, M. T. (1973) Secondary ion emission mass spectra of some polymers. *Khim. Vys. Energ.* **9**, 380–383 (in Russian).
88. Tantsyrev, G. D., Lkeimenov, N. A., Povolotskaja, M. I., and Bravaja, N. M. (1976) Determination of F-polymer microsctructure by the method of secondary emission mass spectrometry (in Russian). *Vysokomol. Soedin.* **A18**, 2218–2222.
89. Tantsyrev, G. D., Povolotskaja, M. I., and Kleimenov, N. A. The use of secondary emission mass spectrometry for determination of orientations of monodimensional glides in macromolecules of F-containing co-polymers (in Russian).
90. Tantsyrev, G. D. and Nikolaev, E. N. (1972) On two mechanisms of water cluster formation at the ion bombardment of the ice film. *Dokl. Akad. Nauk SSSR* **206**, 151–154 (in Russian).
91. Nikolaev, E. N. and Tantsyrev, G. D. (1975) Investigation of charged cluster emission from films composed of a mixture of different isotopic water modifications. *Zh. Tekh. Fiz.* **45**, (2), 400–404 (in Russian).
92. Nikolaev, E. N., Tantsyrev, G. D., and Saraev, V. A. (1976) Secondary water cluster emission of ions of alkali metals. *Zh. Tekh. Fiz.* **46**, 2184–2187 (in Russian).
93. Klöppel, K. D., Bünau, G., and Weyer, K. (1983) Secondary ion mass spectrometry of organic compounds. *Int. J. Mass Spectrom. Ion Phys.* **46**, 463–466.
94. Clampitt, R. (1984) SIMS of solid hydrogen. *Vacuum* **34**, 113.
95. Jonkman, H. T., Michl, J., King, R. N., and Andrade, J. D. (1978) Low-temperature positive secondary ion mass spectrometry of neat and argon-diluted organic solids. *Anal. Chem.* **50**, 2078.
96. Scheiferss, S. M., Verma, S., and Cooks, R. G. (1983) Characterization of organic dyes by secondary ion mass spectrometry. *Anal. Chem.* **55**, 2260–2266.
97. Newbury, D. E. and Simons, D. (1984) The role of standards in secondary ion mass spectrometry. In: *Secondary ion mass spectrometry, SIMS IV*, Springer, Berlin, pp. 101–106.
98. Yamaguchi, N., Honma, y., Kashiwakura, J., and Koike, K. (1984) SIMS quantitative analysis of gallium in silicon by using ion-implanted samples for standards. *Secondary ion mass spectrometry, SIMS IV*, Springer, Berlin, pp. 110–112.
99. Rüdenauer, F. G. Riedel, M., Beske, H. E. *et al.* (1984) An inter-laboratory cross calibration experiment for SIMS. *Quantif. OEFZS Ber.*, NA0561, 27p.

Chapter 4
In-depth analysis

4.1. Technique of SIMS: in-depth analysis

SIMS is primarily a technique for depth profiling. It differs from all other methods in the respect that in-depth penetration and the formation of a useful signal are accomplished in a single easily controllable process. Auger electron spectroscopy combined with ion sputtering may be considered as a rival method, but that technique is in many respects inferior to SIMS [1].

As has been mentioned above, SIMS is based on the analysis of sputtered ion species which represent the initial composition of an object under study. This method is additionally characterized by an exceptionally high relative and absolute sensitivity and a wide range of thickness in the analysed layers. All this makes SIMS indispensable for many applications where other methods fail.

A lot of works [2–27] deal with SIMS potentialities and practical applications for in-depth analysis. At present we are already in a position to judge on its advantages and drawbacks. Continuous measurement of ion peak intensities during sputtering of the surface by an ion beam may be successfully used to study concentration distributions in diffusion and implanted profiles, surface layers and thin films. Such measurements are quite simple from the methodical point of view if the direct proportionality is maintained between the amount of the analysed impurity and the ion yield in the concentration range under study. Such proportionality exists in many important cases and is practically always true for small concentrations (less than 1%) of substitutional and interstitial impurities.

In-depth analysis of high accuracy is possible for layers and films whose thickness does not exceed several hundred nanometres. Otherwise the time of measurement becomes too long and the effect of crater walls is significant. General accuracy of analysis is characterized by the precision with which the coordinate of the layer, whose concentration is to be measured, is determined, that is by in-depth resolution and by the accuracy of concentration measurements.

In-depth resolution depends on the sputtering rate, which in its turn depends on the density of the primary ion current (the primary ion energy is usually kept at a level of 6–8 keV) [28]. When the surface sputtering yield, S, is known, the rate of surface regression into the bulk, V_s, may be found from

$$V_s = j_0 SA_m/e\rho \tag{4.1}$$

Here j_0 is the current density of primary ions on the target, in A cm^{-2}; $e = 1.59 \times 10^{-19}$ K and is the charge of an electron; $m = 1.66 \times 10^{-24}$ g; ρ is the target density, in g cm^{-3}; and A is the target atomic mass.

If in this formula j_0 is measured in mA cm^{-2}, then the sputtering rate (nm c^{-1}) will be equal to

$$V_s = 0.1045 \; \frac{j_0 SA}{@} \tag{4.2}$$

Table 4.1 shows sputtering rates for some materials, calculated from Equation (4.2) for a current density of 1 mA cm^{-2} and a primary ion energy of 8 keV (the data are given in decreasing order). It may be seen that if the secondary ion current is measured with a time constant of about 0.5–1.0 s, an in-depth resolution of the order of 1–2.5 nm can be easily obtained, which is the physical limit of what may be practically realized.

Another important parameter that restricts in-depth resolution is the sensitivity limit of analysis, especially when low concentrations of impurity are to be detected. In this case a target must be sputtered to a certain depth before a sufficient signal can be accumulated at the ion detector output. The ratio between the sputtered volume $V = \Delta Za$ and the number of ions n of relevant impurities with atomic concentration C detected by the mass spectrometer ion detector, is given as

$$n = CNa \, \Delta Z R_i K_t \tag{4.3}$$

where N is the matrix atomic density, in atom cm^{-3}; a is the area under bombardment, cm^2; ΔZ is the layer thickness; R_i is the ionization efficiency of the ith impurity; and K_t is the general transmission in the secondary ion analysis channel from a sample to the detector.

Due to counting statistics, $10^4/p^2$ ions must be detected to provide an accurate concentration measurement. Let $p = 10$ per cent. Then $n = 100$ ions. Taking $N = 6 \times 10^{22}$cm^{-3} as a mean value, the layer thickness to be sputtered in order to provide an accuracy of $p\%$ may be written as

$$\Delta Z \approx 1.6 \times 10^{-21}/aR_iK_tC_{min} \tag{4.4}$$

Table 4.1
Rates of sputtering for various materials (V_s)

Element	V_s, nm s^{-1}	Element	V_s, nm s^{-1}	Element	V_s, nm s^{-1}
C (graphite)	5.70	Ni	3.35	Ti	2.45
Pd	4.90	Fe	3.25	Si	2.35
Be	4.75	Co	3.20	Nb	2.14
Al	4.30	Cr	3.10	W	2.05
V	3.80	Mo	2.70	Ta	1.70
Cu	3.80	Zr	2.70		

For an analysed area 1×1 mm^2, an ionization efficiency $R^+ = 10^{-3}$, instrument transmission $K_t = 10^{-2}$, and $C_{min} = 10^{-6}$, one obtains $\Delta Z = 16$ nm, which characterizes the in-depth resolution limit under given conditions.

When a solid surface is bombarded, the impurity atoms are partially knocked into the target bulk. As a consequence, 'recoil tails' appear in the concentration distribution curves. Calculations have shown that in the energy range up to 50 keV, such 'tails' result from multiple low-energy recoils (0.1–0.2 E_0). Direct recoil of impurity atoms at the collision of the incident particle is much less probable. It has been found that these effects are of no practical consequence for measurements if primary ion energy is below 10 keV [29].

Errors in thickness measurement at initial stages of sputtering may be caused by variation in the sputtering yield. The latter varies until the steady-state concentration of implanted primary ions is reached. This steady-state concentration may be estimated as $C_p \sim 1/S_p$, where S_p is the equilibrium sputtering coefficient. When the target is bombarded by the beam with current density j_0, primary ions penetrate into the lattice. Their concentration distribution may be approximated by Gaussian

$$dC(x') = \frac{0.4}{\Delta\lambda}\left[\exp-\frac{(x'-\lambda)^2}{2\Delta\lambda^2}\right]\frac{j_0}{e}\,dt \qquad (4.5)$$

Here x' is the depth stationary coordinate; λ is the maximum range of the primary ion penetration; $\Delta\lambda$ is a root mean square deviation of the projected range; and $j_0\,dt/e$ is the ion dose.

The λ and $\Delta\lambda$ values depend on the primary ion energy. If the target is sputtered at the rate V_s the surface will regress into the bulk, and its instaneous position x will be determined as $x = x' - V_s t$. Then

$$dC(x) = \frac{0.4j_0}{e\Delta\lambda}\exp\left[-\frac{(x+V_s l - \lambda)}{2\Delta\lambda^2}\right]dt \qquad (4.6)$$

Integration of this expression allows calculation of the primary ion concentration as a function of time and depth:

$$C(x) = \int_0^t \frac{0.4j_0}{e\Delta\lambda}\exp\left[-\frac{(x+V_s t-\lambda)^2}{2\Delta\lambda^2}\right]dt$$

or, using the probability integral,

$$C(x,t) = \frac{j_0}{2eV_s}\left[\text{erf}\frac{x-\lambda+V_s t}{\sqrt{2}\Delta\lambda}-\text{erf}\left(\frac{x\lambda}{\sqrt{2}\Delta\lambda}\right)\right] \qquad (4.7)$$

It takes an interval t_0 to sputter the layer $V_s t_0 = \lambda$ and to reach the steady-state such that

$$C(x,t_0) = \frac{j_0}{2eV_s}\left[\text{erf}\left(\frac{x}{\sqrt{2}\Delta\lambda}\right)-\text{erf}\left(\frac{x-\lambda}{\sqrt{2}\Delta\lambda}\right)\right] \qquad (4.8)$$

The primary ion surface concentration will then amount to

$$C(0) = \frac{j_0}{2eV_s} \, \text{erf} \left(\frac{\lambda}{\sqrt{2}\Delta\lambda} \right) \qquad (4.9)$$

Hence, the lower the surface concentration of primary ions, the less the sputtering rate changes. This may be realized through the compromise of decreasing the primary ion beam current density and energy. If we assume V_s to be constant for a given j_0 the apparent thickness of the removed layer $V_s t = x^*$ may be found instead of the true x. For $x \geqslant \lambda$ we get

$$x = x^* \, (1 + \delta \, x/x) \, \delta x \qquad (4.10)$$

The x value may be defined from the measurement of an assumed and true depth of impurity found by means of independent measurements. It has been established [28] that $\delta x/x$ does not exceed 2.4% even at the accelerating voltage of Ar^+ ions (50 kV), while for lower energies of the beam it may be below 1%.

The most important error sources in evaluation of $C(x)$ are the formation of some ion etching topography and modification of crater bottom flatness, due to the fact that the current density is non-uniform across the primary ion beam.

The current density profile for the focused ion beam is described by a Gaussian of the form

$$j = j_0 \exp \left[-(r/r_s)^2 \right] \qquad (4.11)$$

where j_0 is the maximum current density in the centre and r_s is the beam nominal radius. Total current equals in this case $j_0 r_s^2 \pi$, i.e. the ion beam with the same total current but with constant density j_0 might have the radius r_s. It is quite clear that the focused ion beam which irradiates the sample surface will produce there a crater of the same profile. Hence the dependence of the secondary ion current intensity should be written as

$$I = I_0 \, (x,y) \, j_i C_i (x,y,z) \qquad (4.12)$$

This implies that when the primary current density is constant across the beam section, the secondary ion current variation is proportional to the variation in the concentration with depth. However, when secondary ions are collected from the total surface of the crater (walls and bottom) as it becomes deeper, no useful information about in-depth composition can be obtained. Various ways have been proposed to solve this problem [16, 30]. The simplest one is secondary ion collection from a limited area on the crater bottom [Fig. 4.1(a)]. The limits can be set by means of a collimator. Since in this case the accelerating field is not used, the efficiency of secondary ion collection is low. However it may be increased when a selected area is projected as an ion image on the selector aperture plane [Fig. 4.1(b)] followed by the mass analyser.

Local in-depth analysis for a small depth can be performed by collimating the primary beam in such a way that a probe of uniform current density is

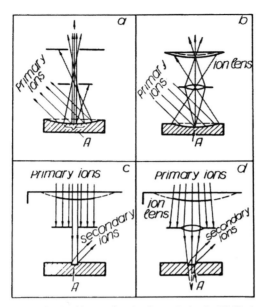

Figure 4.1 Instrumental methods of improving depth resolution in SIMS. A — area under study.

obtained [Fig. 4.1(c)]. Higher current density, and hence shorter sputtering time, are possible if such a beam with uniform current density is weakly focused [Fig. 4.1(d)] [12]. But the use of the two latter methods does not completely exclude the effect of crater walls on the depth resolution. The remedy may be found in the so-called 'electron aperture' used in ion microprobes [23] (Fig. 4.2). The focused primary beam is scanned over the surface and produces a raster. When the interline spacing equals r_s, the non-homogeneity of the current density due to overlapping of traces amounts to less than 1%, and ensures the homogeneity of sputtering and a flat bottom to the crater. The effect of the crater walls can be avoided in this case by switching off the detector when the ion beam is on the raster edges. As a

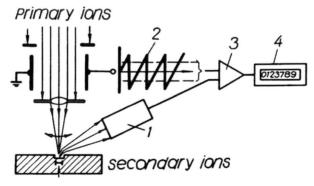

Figure 4.2 Diagram of the 'electron aperture'. (1) Mass spectrometer; (2) deflecting voltage; (3) amplifier; (4) scaler.

result, only ions from the central part of the raster reach the output. An alternative solution is a spiral scanning by two sinusoidal voltages, phase shifted by 90° (Fig. 4.3) and applied to the deflecting plates. When the amplitude of these voltages, V_d, is increased according to $V_d \sim V_t$, the spiral radius is also increased, and the crater has a circular flat bottom. The detector is switched off on the last outer winding of the spiral and thus electronic blending is provided, excluding the effect of the crater walls. Hence the problems of high in-depth resolution may be successfully solved by purely instrumental means.

Physical factors can be also controlled to some extent. In particular, penetration of primary ions into the target can be decreased by lowering their energy or using heavy primary ions. But at lower ion energy the sputtering and secondary ion yields are decreased, which is undesirable when a high sensitivity of analysis is required. Very low energies of primary ions (up to 500 eV) enhance the effects of differential sputtering and surface migrations of target atoms. At a given primary ion energy and mass the incident angle increase leads to a smaller projected mean range and to a larger yield of sputtered atoms almost in proportion to θ (for $\theta < 60°$). But simultaneously the primary current density drops in proportion to $\cos\theta$, so that the etching rate remains practically unchanged, though the decrease in the mean projected range improves in-depth resolution.

The simplest way to estimate the depth resolution is to measure the speed with which the peaks of definite secondary ions disappear when a reference monoelemental thin film of high structural perfection is completely sputtered. The slope of a final drop in the curve $I_i^+(t)$, e.g. from 90 to 10% of the maximum signal, serves as a measure of depth resolution. In such a way it is possible to choose sputtering parameters which ensure a resolution of about 2–4% of the layer thickness [28].

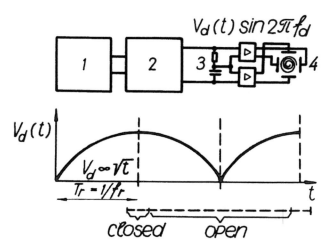

Figure 4.3 Scheme of electron aperture with spiral scanning. (1) Generator of sinusoidal voltage; (2) amplitude scanning; (3) 90° phase shifter; (4) deflecting system.

4.2. Study of thin films (vacuum condensates)

Multicomponent metallic and semiconductor thin films of stoichiometric or alternate composition find steadily wider application in microelectronic devices. Element distribution through the film thickness, and the degree to which initial substance composition is reproduced in the film, are important parameters for thin film deposition and the study of their properties. In this respect SIMS is the most promising method, since it combines high sensitivity and depth resolution on the level of several nm.

The systematic study of SIMS potentials and their application to thin film production has been carried out in a number of works [3–5, 14–16, 19].

The structure, composition, and properties of even one-component films produced by thermal evaporation are known to be determined by condensation conditions and depend on a number of factors, pressure and composition of residual atmosphere being the most important ones. During the condensation process the substance being evaporated can interact with residual gases. In addition, the freshly deposited film surface actively sorbs the gases, and sometimes chemical compounds can be formed. The study of these processes is of great practical interest for evaluation of the physical properties of such films. SIMS potentials in the solution of these and related problems are illustrated below.

4.2.1. Surface Layers of Vacuum Condensates

As has been mentioned above, when a bulk target is bombarded with Ar^+ ions, the emission intensity is at first sharply increased (surface peaks) but with further sputtering becomes weaker and attains a stable value. It has been found from the study of monoelemental films of Al, Ti, Mn, Ni, Cu, Sn, Ta, and Au bombarded with Ar^+, He^+ and O_2^+ ions (8 keV, current density about 0.5 mA cm^{-2}) that surface peaks for films are similar to those observed in kinetic curves at sputtering of massive targets. Table 4.2 gives the relative height of surface peaks for Me^+ ions from investigated films bombarded with Ar^+, He^+ and O_2^+ ions. It may be seen that, as in the case of a massive sample, the surface peak height is a function of the target atomic number and depends on the primary ion nature. The most intense peaks are typically for metals with high oxygen affinity. This is reasonable since in this case the peaks reflect the ability of the metal to oxidize. In this connection it is interesting to note that the bombardment with O_2^+ ions leads to a significant decrease in the

Table 4.2
Relative height (%) of surface peaks
$[h = (I_{peak}^+ - I_{st}^+/I_{st}^+)]$

Ion	Films							
	Al	Ti	Mn	Ni	Cu	Sn	Ta	Au
Ar^+	1500	3000	2200	400	2100	900	1700	30
He^+	0	91	880	800	550	800	300	20
O^+	0	4	1100	90	60	0	50	0

surface peak magnitude, due to a general secondary ion current increase as a result of oxygen ion action on the process of surface oxidation ('chemical SIE').

Figure 4.4 shows how the secondary ion current of $^{181}Ta^+$ depends on the duration of Ta film sputtering with O_2^+ ions (oxygen partial pressure in the target region 4×10^{-3} and 8×10^{-4} Pa) and with Ar^+ ions. It may be seen that in case of oxygen with a high partial pressure, the surface peak is absent while the emission current is high. These data indicate that one of the main reasons why initial spikes appear at the bombardment with inert gas ions is oxidation of topmost film layers. Hence the peak height characterizes the degree of oxidation, while the interval during which the current is stable, τ_{st}, defines the thickness of the oxide layer.

Figure 4.5 shows the results of surface peak investigation for Mn, Ni, Ti, and Cu films prepared by vacuum evaporation of the corresponding bulk metals onto a substrate held at 300°C. The dependence on film temperature at the moment of chamber opening is shown. When the vacuum chamber is opened at a high substrate temperature (100–300°C) the film surface is strongly oxidized. Thus, for example, the spike height for $^{58}Ni^+$ ion current at the substrate temperatures of 20, 200, and 300°C makes up 400, 990, and 1300%, respectively. Films exposed to air at high substrate temperature are oxidized deeper, as may be judged from the longer time taken to reach emission stability.

Hence the study of surface peak nature presents special interest because it gives unique information about the oxidation of the vacuum condensate surface. The kinetics of this process may also be studied. In this case it is useful to record and then compare the curves $I_i^+ = f(\tau)$ for both Me_i^+ ions and $^{16}O^{\pm}$ ions.

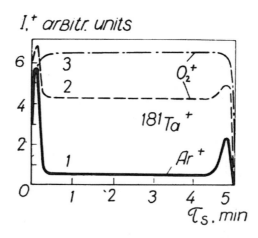

Figure 4.4 $^{181}Ta^+$ ion currents vs time of sputtering of Ta film by different primary ions (1) Ar^+; (2) O_2^+ with oxygen residual pressure 8×10^{-4}Pa; (3) O_2^+ with oxygen residual pressure 4×10^{-3}Pa.

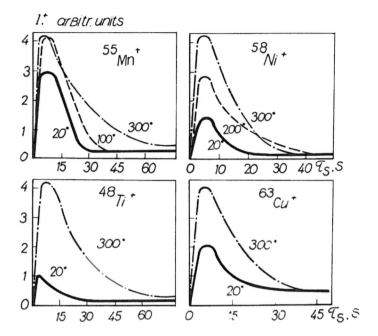

I^+_i arBitr. units

Figure 4.5 Kinetic curves of secondary ions at initial stages of sputtering of Mn, Ni, and Cu films, depending on temperature at which the deposition chamber was opened (numbers near curves — 8 keV Ar^+ primary ions.

4.2.2. Investigation of the Film – Substrate Interface

A significant current increase is usually observed for secondary ions originating from the substrate (this increase will be labeled below as 'the final peak') in the moment when Ar^+ and He^+ ions bombarding the films reach the layers adjacent to a dielectric substrate. The relative height of the final peak h makes up 100–700%, depending on the film.

Figure 4.6 gives as an example the dependence of $^{63}Cu^+$ ion current on the sitall deposited film thickness under Ar^+, He^+, O_2^+ bombardment.

In the case of Ar^+ the I_i^+ (τ) curves for negative ions $^{16}O^-$, $^{26}C_2H_2^-$, and $^{35}Cl^-$ also have surface and final spikes, the peaks for $^{26}C_2H_2^-$ and $^{35}Cl^-$ ions being much narrower than for Me^+ and O^- ions and somewhat shifted along the time axis with respect to the latter [Fig. 4.6(b)].

The freshly deposited film actively adsorbs gas molecules so that the first inner layers, which adsorb most residual gases turn out to be more strongly contaminated. Evolution of condensation latent heat favours formation of chemical compounds (in particular, oxides) in the layer adjacent to the substrate.

Hence the observed increase of Me^+ and O^- currents at the end of the sputtering is predominantly due to presence of an oxide layer on the substrate. The final peaks of $C_2H_2^-$ and Cl^- ions are perhaps connected with contaminations adsorbed on substrate prior to film deposition. It may thus be

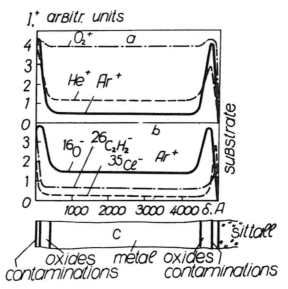

Figure 4.6 Depth profiling of Cu film. (a) Emission of $^{63}Cu^+$ ions at Ar^+, He^+, and O_2^+ ion bombardment. (b) Emission of $^{16}O^-$, $^{26}C_2H_2^-$, $^{35}Cl^-$ at Ar^+ bombardment. (c) Diagram of interfaces in the film.

concluded that for the systems 'vacuum–film' and 'film–substrate' the interfaces are similar to each other in their physico-chemical state. Analysis of final peaks may give useful information about the regularities of the vacuum condensate's adhesion to substrates of different materials.

4.2.3. In-depth Complete Analysis

The potential of SIMS may be illustrated by the results obtained for a series of multicomponent films. For instance, the system Ni/AuGe was prepared by the following process. The AuGe eutectic alloy was evaporated onto a GaAs crystal so that a film of thickness $\sim$ 500 nm was formed. Ni film (thickness about 50 nm) was then deposited. The system obtained was annealed at 480°C for different periods (15 s, 2 min, and 5 min). Thus films with different parameters were produced. In-depth analysis has shown that variation in heat treatment causes changes in component distribution through the film thickness. Figure 4.7 represents the depth distribution of Ni, Au, Ge, and Ga in the as-deposited and annealed (15 s) films. It can be seen that the annealing causes Ni diffusion into the AuGe layer and Ga diffusion into the film, while Au penetrates to the system surfaces.

The data on systematic study of condensate formation in case of thermal evaporation of copper-based Cu–Mn–Ni–Ti alloys are also of interest [8]. Figure 4.8 represents the data on component distribution across the film thickness for different rates of deposition. Since the secondary ion current is in this case directly connected with the concentration of a given element, at least qualitative judgement is possible on concentrational distribution, from distribution profiles. As can be seen, evaporation of Cu–Mn–Ni–Ti alloy

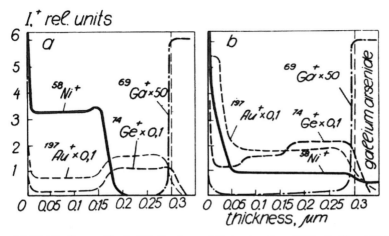

Figure 4.7 Distribution profiles for Ni, Ge, Au, and Ga in a two-layer Ni/AuGe film at GaAs. (a) Freshly deposited film. (b) Annealed at 480°C for 15 s (Ar⁺, 6 keV).

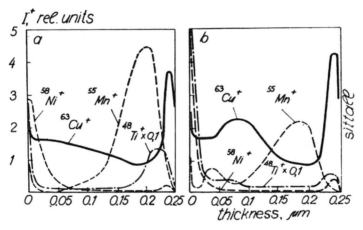

Figure 4.8 Distribution profiles for components in the films prepared by vapour deposition of copper alloy containing 5% Ni, 2% Mn, and 0.1% Ti with deposition rates of (a) 18, and (b) 6.8 nm s⁻¹ 0Ar⁺, 6 keV.

produces a film with a non-uniform element depth distribution: the surface is enriched with Ni and Ti, while Mn is accumulated near the substrate. Copper creates the condensate matrix. In order to exclude the effect of surface and interface peaks the films were bombarded by O_2^+ ions at high partial oxygen pressure. It could be concluded from the results obtained that during alloy evaporation a definite fractionation occurs because of significant differences in the thermodynamic parameters of the components.

It follows from Figure 4.8 that at low deposition rates Mn and Ti are practically completely evaporated during the initial period. With the higher deposition rate, Mn appears within the film bulk and at the surface. Ti and Ni are evaporated mainly at a final stage and cover the film surface.

Secondary ion emission is known to be enhanced in chemical compounds as compared to emission from pure metals. This fact may be used for detection of new phases formed in the films as also for the analysis of their distribution across the film thickness. As an example, Fig. 4.9 shows the dependence of secondary ion currents for components in the film prepared by evaporation of a Cu (2.0%) – Cr (4.6%) – Al alloy. Consideration of these curves and relevant data for bulk samples of alloys with different concentrations leads to the conclusion that a σ-phase (~35% Al) is produced in films in about ⅘ of their thickness. The surface layer is formed by pure copper. Chromium does not participate in phase formation and almost all its content may be found in the substrate. This picture is supported by a similar distribution obtained in the case when the film is sputtered by O_2^+ ions.

4.2.4. Films of Chemical Compounds

Films of various chemical compounds (CdTe, CdP, PbAgAuIr, Dy_2O_3, $Nd(OH)_3$, $Me^{IIIB}N$, Nb_3Sn, etc.) prepared by thermal evaporation and cathode sputtering have been studied. Figure 4.10 shows component distribution in CdP_2 film: the film after deposition had been annealed in Cd vapour at different partial pressures. It can be seen that at higher Cd vapour pressures the film surface is more saturated with Cd while P is pushed down towards the substrate.

In-depth comparative analysis of nitride films of metals from Group III B and data from other methods showed that formation of a TiN matrix is possible in films; this is impossible in the bulk state.

SIMS can provide additional information on the reproducibility of the stoichiometry of an initially stoichiometric substance in a corresponding film; also on variations in the stoichiometry with depth. Let $A_mB_n(C, 1-C)$ be the initial compound. The stoichiometry of the film of unknown composition $A_xB_y(C, 1-C)$ may be easily found (with the use of measured secondary ion currents) from the relation

$$C_A = C_A/(RC_B + C_A) \tag{4.13}$$

where $R = I_A^+(I_B^+)'/I_B^+(I_A^+)'$.

For a ternary system (A–B–C) the following relations may be used:

$$\frac{I_A^+I_B^+}{(I_A^+)'(I_B^+)'} = \frac{C_AC_B'}{C_BC_A'} = R, \quad \frac{I_A^+I_C^+}{(I_A^+)'(I_C)'} = \frac{C_AC_C'}{C_CC_A'} = S \tag{4.14}$$

$$C_A' + C_B' + C_C' = 1$$

$$C_A' = C_A/(C_A + C_B R + C_C S), \quad C_B' = C_B R/(C_A + C_B R + C_B S)$$

The simplest criterion for preservation of stoichiometry in the film is the equality of ion current ratios for components in the bulk sample and in the corresponding film [4].

4.2.5 Gas Impurity Analysis

SIMS is also applicable to the study of the concentration distribution of gaseous impurities across the thickness of thin films. It has turned out that for some elements (O, Cl), increased sensitivity is achieved using negative secondary ions. We studied, for example, the distribution of H^+, H_2^+, $O^{+(-)}$ in Pb films [4], C^-, O^-, Cl^-, H^+ in Nb films [9], and N^+, H^+, $O^{+(-)}$ in Ge and TiN films. It has been found in particular that residual gases strongly affect the electrophysical properties of vacuum condensates.

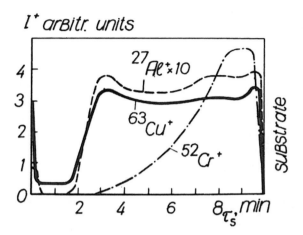

Figure 4.9 Distribution of components in a film prepared by evaporation of Cu–Al–Cr alloy.

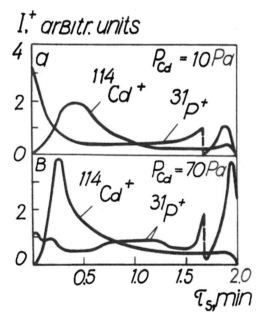

Figure 4.10 Dsitribution of components in CdP_2 film annealed in cadmium vapour at partial pressure of (a) 10 Pa, and (b) 70 Pa. He^+, 8 keV.

4.3. In-depth analysis of implanted profiles

Production of thin surface and subsurface layers containing controlled amounts of impurities of a definite type is of utmost importance in the technology of integrated circuit production. Such layers can be most efficiently produced by ion implantation, while the study of the concentration distribution profiles obtained may be successfully carried out using SIMS, whose success in this application alone has stimulated the acknowledgement of the method and has led to its wide use in scientific and industrial investigations.

In order to solve the problems of in-depth analysis of implantation profiles all the SIMS potentialities should be mobilized to provide the maximum concentration sensitivity and, at the same time, a high in-depth resolution. These requirements are contradictory. For detection of concentrations at the level of 10^{15}–10^{17} atoms cm^{-3} (10^{-6}–10^{-5} atom%) with a depth resolution of 2-3 nm the instrument must have maximum transmission of the secondary ion analysis and detection channel. Narrow profiles of shallow depth can be analyzed readily enough, since in SIMS they correspond to initial sputtering stages where the crater is not yet deep, and the amount of target material sputter deposited on the chamber walls and on the electrodes of the secondary ion optics is not large. Difficulties arise when analysing distribution 'tails'. Figure 4.11 shows schematically the contributions of various sources of uncertainty to the depth analysis of implantation profiles. The important role of instrumental factors in depth analysis is evident.

The typical problem to which a number of works were devoted is the analysis of Si or SiO$_2$-implanted boron profiles. Figure 4.12 gives examples of profiles for implantations at different energies.

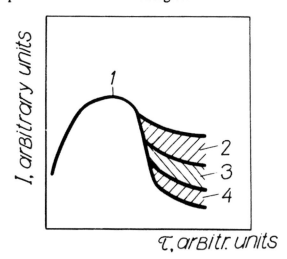

Figure 4.11 Effect of various sources of inaccuracy in the depth analysis of implantation profiles. (1) True profile; (2) contribution from crater walls; (3) contribution from secondary sputter deposition and memory effects; (4) contribution from superposition in mass spectrum. τ — time of sputtering; I — secondary ion current.

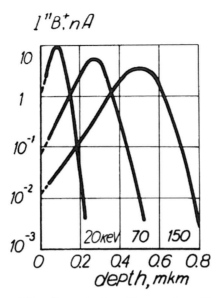

Figure 4.12 Distribution profiles of boron implanted in amorphous silicon (energy 20, 70, and 150 keV with doses of $10^{15}cm^{-2}$). SIMS data, Ar^+, 10 keV.

Detailed information has now been accumulated about the parameters of implantation of different elements into simple and layered structures [13, 25, 27]. The depth analysis has been performed with simultaneous electrical study of MOS structures [26]. Distribution profiles of such important dopants as arsenic and phosphor, which are implanted into Si or SiO_2, are usually difficult to study because of mass spectral interference produced by superposition of cluster ions from the target material. As a result, the sensitivity of the analysis is lowered by several orders of magnitude. This problem arises with bombardment by oxygen primary ions or inert gas ions in the presence of oxygen used to increase the measurement sensitivity. In such cases the spectra contain ions of $M_mO_m^{\pm}$ type oxides with all the possible combinations of m and $n(m,n = 1, 2, 3,...)$. The emission intensity of these ions is so high that it becomes quite impossible to analyse microimpurities when superpositions occur in the mass spectrum. For instance, the analysis of arsenic in Si is hampered by the fact that the $^{75}As^+$ ion has the same mass number as the ion $^{29}Si, ^{30}Si^{16}O^+$. Depending on the width of the energy window in the mass spectrometer, it is possible to get the ion current of $^{29}Si^{30}Si^{16}O^+$ similar to that of $^{75}As^+$ ions at an As concentration of $(1–2) \times 10^{19}$ atoms cm^{-3}. The interference between $^{31}P^+$ and $^{30}Si^1H^+$ ions is similar. These ions are produced by water vapour adsorption from residual atmosphere on the sample surface. At a residual pressure of even $4 \times 10^{-7}Pa$ the intensity of $^{30}Si^1H^+$ emission corresponds to a P concentration of about 2×10^{18} atom cm^{-3} [22].

In order to separate relevant duplets, the mass spectral resolution must be of the order of $M/\Delta M = 4000$ for P and about 3200 for As. This is clearly impossible in simple SIMS instruments, and hence other means of analysis should be looked for. For instance, the currents of $^{31}P^{16}O^+$ or $^{75}As^{16}O^-$ might be measured instead of $^{31}P^+$ or $^{75}As^+$. This ensures sensitivity sufficient for concentration analysis at the level of about 10^{18} atom cm^{-3} [22]. Sensitivity may be improved by using different characters of monatomic and cluster ion energy distributions. As has been shown above, clusters containing more atoms are characterized by a shorter and lower high energy 'tail' in the distribution. Therefore by setting the energy 'window' in the elevated energy range it is possible to suppress cluster ions fairly efficiently. In the DIDA instrument (see Chapter 2) the energy window can be set by applying corresponding potential to the target so that the transmission of the analysis and detection channel for ions of different species can be monitored over a wide range. The efficiency of this method is illustrated in Fig. 4.13 [22]. As may be seen, the quadrupole maximum transmission corresponds to a target potential of about 2V. However in this case the bombardment of the Si target by O_2^+ ions with 12 keV energy allows the detections concentrations not lower than 10^{19} atom cm^{-3} even if the Si_2O^+ background has been subtracted. When the system energy transmission corresponds to the target potential of 25 V, the current ratio of As^+ and Si_2O^+ ions is increased by two orders of magnitude which indicates a corresponding gain in the sensitivity for As identification.

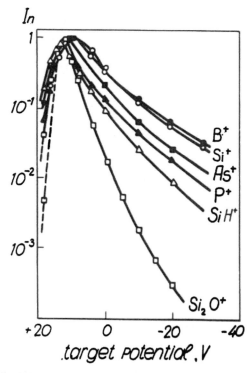

Figure 4.13 Normalized distribution curves for secondary ions ejected from an Si target homogeneously doped with B, P, or As (I_N^- normalized intensity) From [22].

Ion energy discrimination is less successful for P where the 'tail' of $^{30}Si^1H^+$ energy distribution does not fall as fast as in Si_2O^+ case. But even for the P depth analysis in Si the sensitivity can be markedly improved: the detection limit may reach several digits of 10^{17} atom cm^{-3}.

In-depth analysis of concentration distributions of inert gases in solids implanted by artificial or natural bombardment, as for example in the case of sun the wind action on lunar regolith particles, is of significant interest. In such cases the use of a standard depth analysis by means of SIMS does not give reliable results because ionization efficiency is very low for implanted inert gases. Measurements have shown [31] that the ionization probability in SIMS is about 10^{-10} for helium, 10^{-6} for argon. Therefore, when a standard technique is used the minimum concentration at the level of $10^{-2}\%$ can be determined, this being approximately two orders of magnitude higher than the Ar maximum surface concentration in samples of natural origin, e.g. in the moon regolith. The situation is still worse for helium and neon. The difficulty may be overcome if the release of implanted gases due to ion bombardment and subsequent ionization of the released gases by electron impact is combined with the analysis of ions obtained from the mass spectrometer. The diagram of such an arrangement, which has been given the name of the 'gas ion probe' [31], is shown in Fig. 4.14. The ion beam (1) is bombarding the target (2) placed in the chamber (4) filled with a neutral gas for calibration. Sputtered atoms and molecules are thermalized in the course of interaction with the chamber walls and are transfered into the chamber (5) where the electron impact causes ionization. The ions obtained are then directed to the mass spectrometer.

The accuracy of the results depends on the correct choice of chamber dimensions and aperture. This choice is based on the following arguments. Gas flow from the target (6) into the ionization chamber must significantly exceed its leakage from orifices intended for the primary beam and for admission of the calibration gas. Therefore the primary ions are admitted through an aperture 200 μm in diam. while the reference gas enters through aperture (3), which has a diameter of 300 μm. The chambers for bombardment and ionization are connected to each other via an orifice 3 mm in diam. Hence only about 1.5% of the gas extracted from the sample is lost through the small orifices.

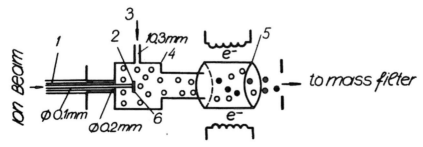

Figure 4.14 Diagram of the gas ion probe.

On the other hand, the time # during which gases from the target remain in the chamber (4) must be short enough not to reduce the depth resolution of concentration measurements. The larger the ratio between the exit orifice area and the chamber volume V, the shorter is the time τ. At $V = 1$ cm^3 $\tau \sim$ 10^{-3} s for ^{40}Ar$^+$. This implies that ion currents measured at the mass spectrometer output must follow gas evolution from the target with a time constant of the order of 1 ms, i.e. the depth resolution depends on the sputtering process itself. Figure 4.15 gives an example of gas depth analysis. Distribution of ^{4}He was measured at the surface of lunar glass brought to Earth by the Apollo 17 mission. The depth resolution and concentration sensitivity are seen to be sufficient to analyse in detail the character of the gas distribution in various objects.

Depth analysis by SIMS has also been widely used to solve different problems in the field of metal physics and material science. A lot of works deal with the study of lithium vapour interaction with single- and polycrystal-line refractory metals, in particular, with W and Mo [32]. In all cases SIMS has proved to be extremely efficient and has led to some new results.

In-depth analysis applications of SIMS are not limited to those described in this chapter. Its use for the study of grain boundary segregation appears to be promising: not for a common case when the boundary is opened and the composition of fracture surface has to be investigated, but for the case when a closed boundary must be studied. To realize this idea a suitable bicrystal is chosen. The crystal is cut away and polished in such a way that its outer surface is parallel to the boundary at the distance of some micrometres. Then the remaining material is sputtered by an ion beam, which crosses the boundary and allows definition of the composition in the vicinity of the boundary. In this case the interaction with the residual gas atmosphere has no effect, nor do mechanical interactions and structure distortions unavoidable at the boundary disclosure at sample fracturing.

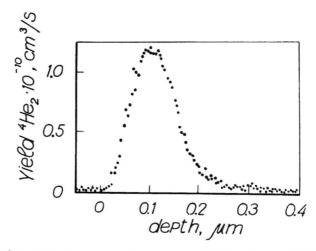

Figure 4.15 ^{4}He$_2$ distribution curve for lunar glass; glass 76501, sputtering rate 1.0 nm s^{-1} [31].

4.4. Spatially multidimensional SIMS analysis

It was recognized at an early stage of SIMS development that the lateral resolution capability of the scanning ion microprobe and the ion mass spectral microscope combined with the capability for controlled removal of surface layers by the sputtering action of primary beams make SIMS an inherently three–dimensional method of microanalysis. In other words, it is possible to obtain information on the composition of a microvolume (voxel) of a solid as a function of the x, y, z coordinates of that voxel.

First reports on three-dimensional SIMS (see Refs 2 and 3 in [33]) describes its realization by photographically recording a time-series of concentration ion images of the sample area, corresponding to different depths of sputtering. However, analysis and processing of these images was cumbersome and time-consuming, and the information obtained was of a qualitative nature.

A large step forward in the problem of three-dimensional analysis became possible only when on-line digital image acquisition and processing techniques were introduced into SIMS analysis [34–36]. A major problem in multidimensional analysis is quantitation. Although collection of a three-dimensional data set is basically equivalent to a large number of individual quantitive point analyses, computation time becomes a major limiting factor which prohibits the use of extensive accurate correction algorithms. Other problems are obtaining correct spatial registration of the analysed volume elements, and the progressive degradation of the sample due to gradual relocation of sample atoms, particularly during deep sputter penetration (thus the sample composition at any instant is not identical to its composition before sputtering). All these problems impose practical limits on the spatial and compositional accuracy of multidimensional SIMS analysis.

Obtaining of two-dimensional elemental distributions in a digital form is a first step to three-dimensional analysis. In the case of the ion microprobe this was first achieved by using a scanning digitized primary beam. Local ion counts for each voxel were stored on punched paper tape, and these data were then processed and displayed on an off-line computer. Digital image acquisition proved to be more cumbersome in ion microscopes. In this case the photographs of ion images were recorded directly from the screen of a CAMECA SMI–300 microscope using a standard photocamera. The images were then digitized by running negatives through a digitally controlled microdensitometer coupled to the digital data storage unit (magnetic disc) for subsequent processing and display [37].

The first modern system for on-line digital image acquisition, storage, and display was created by Rüdenauer and Steiger [34] in 1981. In their arrangement (Fig. 4.16) the primary beam scan was digitally controlled from a PDP 11/34 minicomputer via a CAMAC interface system. Ion counts from each pixel were stored on-line on magnetic disc. Digitized images were transferred to a digital video storage unit which allowed observation of elemental distributions in different shades of grey or in colour coding on a colour TV monitor. Thus immediate feedback between experiment and displayed images became possible.

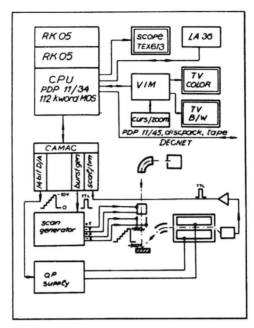

Figure 4.16 Digital image acquisition, storage, and display system for scanning ion microprobe [34].

Similar systems were later installed in other scanning ion microprobes [38]. An important feature of these systems is their quantitative nature in the sense that a number of secondary ions is obtained for every point of the image immediately, without corrections for the non-linear sensitivity of photofilm, or for luminofor background or variation of local sensitivity (all problems found with photographic or TV recording) [37, 39].

Another method of digital image acquisition system for ion microscopes is based on the application of the quantum image detector [40, 41]. This employs a double microchannel plate which converts the ion image into an electron one, and a resistive anode encoder (RAE) digitizes the local position along two coordinates for each ion arriving at the multichannel plate (Fig. 4.17). Three digital and two analog signals are available at the output of this detector: a strobe signal, indicating that an ion has arrived, and two 8 or 10-bit words indicating the X and Y position of the ion. The digitized X/Y information can be used to address a certain location in a fast accumulator memory, increasing the count in the respective memory location by exactly 1. The analog X/Y pulse trains can be fed to the X/Y inputs of a storage oscilloscope, where a bright spot can be marked at the appropriate screen location for each arriving ion. Thus simultaneous observation of a pulse density modulated ion image and permanent digital storage of the same image is possible.

In any of these systems, consecutive scanning of a number of frames – results in a 'stack' of two-dimensional images containing information on the three-dimensional distribution of an element in the mean sampled surface

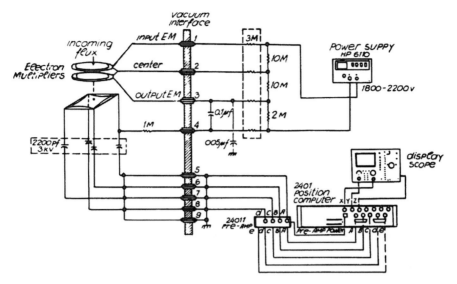

Figure 4.17 Resistive anode encoder (REA) detector for ion images [40].

region of the solid. Since in all the existing probes different elements cannot be detected simultaneously, the three-dimensional multielemental distributions can be recorded only with interlaced elements. It means that the 'stack' contains either two-dimensional images of distribution for a single element or each frame of the stack corresponds to different elements. In this case the mass spectrometer would be repeatedly switched through a 'mass cycle' when one frame is recorded per mass per cycle. Obviously, in the same cycle, frames for different masses are originating from different sample depths and, depending on recording time, may be representative for different incremental layers. Huge amount of data collected in this way and characterizing multidimensional multi-elemental distribution may be sectioned now according to definite features of the data, so that various graphical presentations of images corresponding to different modes of analysis could be displayed.

The problem of displaying these three-dimensional data on two-dimensional display media has been solved in two ways: (a) selection of 'coaxial' and 'trans-axial' slices from three-dimensional data volume and display of these slices as two-dimensional intensity or colour-coded images [39]; (b) display of a series of 'pseudo three-dimensional' (*y*-modulated) images each corresponding to the element distribution at a certain sputter depth [33, 42].

It is obvious that proper selection of data from such three-dimensional image stacks can reproduce all conventional modes of SIMS analysis, i.e. point analysis, line, profile, and image analyses.

Spatial resolution and the sensitivity limits of three-dimensional SIMS analysis are closely interrelated. They are determined by instrumental parameters and by the destructive nature of sputtering and ion formation. The latter sets the physical limit which may be achieved in the ideal instrument, i.e. the one producing any desirable diameter of ion probe, any current density and full transmission and collection of all secondary ions.

As has been indicated in previous sections, if instrumental and physical limits are taken into account, a lower limit to the atomic concentration $C(x)$ of an element X which can be detected with a precision of $p\%$ is given by the formula

$$C(x) = 10^4 / (p^2 \cdot R(x) \cdot T \cdot n \cdot dV) \qquad (4.15)$$

where $R(x)$ is the ionization degree of sputtered X atoms; T is instrumental transmission; dV is sputtered microvolume; and n is the number of atoms per cm^{-3}.

This concentration limit depends on the volume size, dV, but not on its shape. That is why improved lateral resolution may be obtained at the cost of an increase in sputter depth and, conversely, the depth resolution may be improved if a larger area is analysed. If the volume dV is equiaxial, and the spatial resolution is 1 μm, the concentration sensitivity limit C_1 may reach 1 p.p.m. However, in many practical cases of 3-D analysis the depth resolution should be kept as good as possible, say 10 nm. Then Equation (4.15) shows the relationship between the detection limit C_{10} and the smallest possible lateral resolution, d. Figure 4.18 shows the dependence of C_{10} on the lateral resolution d, with the assumption of a useful sensitivity for the element X ($=$ the number of detected X ions/number of sputtered X atoms) of the order of 10^{-3}. These detection limits have been experimentally reached to a lateral resolution as low as 1 μm.

Further improvements of lateral resolution are connected with the development of liquid metal ion sources and their introduction into ion microprobes

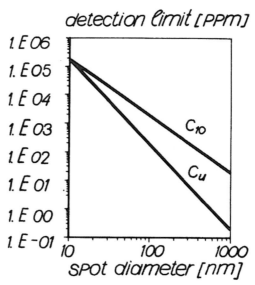

Figure 4.18 Theoretical detection limits for scanning SIMS. A useful sensitivity of about 10^{-3} has been assumed; for explanation see text [33].

(see [25–31] in Chapter 2). True three-dimensional analysis with submicron spatial resolution seems to be possible in this case, and the sensitivity limits in this resolution range can be estimated from Fig. 4.18.

Three-dimensional analysis of real samples is hampered by many factors and artifacts which must be taken into account in order to obtain a true elemental distribution. Local modulation of the image intensity is proportional not only to local element concentration but is affected by surface topography (original or produced by sputtering), energy distribution (different for different elements), matrix effects, and crystallographic contrast, which is caused by the channeling of primary ions and depends on the primary beam incidence angle with respect to local crystal orientation. Most of these factors are taken into account in the standard mode of operation in point SIMS analysis, where known quantification algorithms may in principle be applied at each point to convert the raw data of three-dimensional analysis into true elemental distributions. However, such a correction of a multidimensionl image poses considerable technical difficulties, such as the large amount of computer storage required, and the unsatisfactory low speed of software quantification algorithms on minicomputers, and even on the majority of big machines. When the origin of the analytical signal cannot be correctly located in the x, y, z coordinate system of the sample, the particular voxel is said to be misregistered. Surface topography and local variations in sputtering yield due to local alterations in impact angle, sample composition, or crystallite orientation are the most frequent sources of misregistration of two- and three-dimensional distribution data.

The first approach to local sputter speed evaluation was based on the 'burn-through maps' [36]. In order to obtain such maps a substrate covered with a thin film is sputtered. At those spots in the image where substrate signals appear early in the sputtering process, the sputtering rate must be higher than at spots where the signal appears later. Such a map, in the form of a matrix of correction coefficients, may be used for raw data processing, so that the time axis in profiles can be transformed into a 'true depth axis'.

Most important among the sources of misregistration are atom mixing and relocation of sample atoms along the sample surface. In the course of ion bombardment when the sputtering front attains a certain depth from the original surface, a dynamic equilibrium concentration of some kind is established at this front. But atoms and ions which are sputtered at this stage and contribute to the local SIMS signal do not necessarily occupy this position in the original unsputtered sample. We deal here with sample modification by the measuring process itself. This constitutes the 'uncertainty principle', in sample atom localization. Relocation of atoms in a collision cascade, atom recoil and knock-on mixing can occur across microscopic distances of the order of tens and hundreds of interatomic distances. These effects have been theoretically treated in a number of works and experimentally studied in applications of SIMS and AES. A comprehensive review on this subject has been published by Wittmaack [43]. One important practical conclusion from these calculations and experimental measurements is that very low primary energies (around 1 keV) should be used to reduce the effects of profile broadening. Experimental data [44, 45] show that even in this case the energy-dependent 'tailing' of profiles is observed.

The range of lateral collisional mixing is comparable to that of depth mixing and is too small to produce perceptible effects on lateral resolution for existing microprobes. In the next generation of microprobes, with submicron lateral resolution using liquid metal ion sources, lateral mixing will probably play some part and will have to be corrected for in a proper assessment of lateral elemental distributions.

Relocation of sample atoms across macroscopic distances (hundreds of microns) along the sample surface may occur when a sample atom is sputtered but recondenses again at some other place on the surface after it has travelled some distance in the vacuum or after its reflection from secondary ion optics. Such effects of 'self-contamination' are strongest for topographically structured surfaces with large differences in the height of structural features. This may be true for the analysis of integrated circuits. It has been shown [46] that an Al signal at a distance of 15 μm from an Al step 1 μm high in the integrated circuit amounts to 20% of the signal from the step itself. According to other data [33], lithographically prepared Mo microdots of 200 nm height on an Si wafer, when analysed with a scanning microprobe, manifest an Mo contribution of less than 1% at a distance of 100 μm from a dot. The effect of such cross-contamination in the field of submicron 3-D SIMS analysis is at present not clear but it may pose some problems and limitations for elemental mapping on a submicron scale.

References

1. Saeki, N. and Shimizu, R. (1978) Thickness and in-depth composition profile of altered layer caused on Cu–Ni alloy surface due to preferential sputtering. *Surf. Sci.* **71**, 479–490.
2. Benninghoven, A. (1969) Eine Massenspektrometrische Methode zur Bestimmung von Zerstäubungsrate und Sekundärionenausbeuten Beliebiger Substanzen mit Hilfe dünner Schichten. *Z. Angew. Phys.* **27**, 51–55.
3. Evans, C. A. and Pemsler, J. P. (1970) Analysis of thin films by ion microprobe mass spectrometry. *Anal. Chem.* **42**, 1060–1064.
4. Vasil'ev, M. A., Chenakin, S. P., and Cherepin, V. T. (1974) In-depth analysis of thin films by the secondary ion–ion emission method . In: *Poluchenie i svoistva tonkyh plenok*, Vol. 2, Inst. Problem Mater. AN USSR, Kiev, pp. 16–20 (in Russian).
5. Vasil'ev, M. A., Zaporozhets, I. A., Chenakin, S. P., and Cherepin, V. T. (1974) The use of the mass spectrometer MI–1305 equiped with an ion probe for the in-depth analysis of thin films. In: *Fizikotekhnologischeskie voprosy kibernetiki*, Inst. Kibernetiki AN USSR, Kiev, pp. 56–64 (in Russian).
6. Vasil'ev, M. A., Chenakin, S. P., and Cherepin, V. T. (1976) Peculiarities of the secondary ion–ion emission from the interface between different metals. *Izv. Akad. Nauk SSSR, Ser. Fiz.* **40**, 2571–2574 (in Russian).
7. Vasil'ev, M. A., Kostjuchenko, V. G., Krasjuk, A. D. *et al.* Secondary ion–ion emission coefficients for pure metal thin films. In: *Vzaimodeistvie atomnikh chastits s tverdym telom.* Izd. Khark. Univ., Khar'kov, pp. 102–105 (in Russian).
8. Vasil'ev, M. A., Popov, V. I., Trofimenko, A. D., and Chenakin, S. P. (1977) Regularities in thin film composition formation on the base of multicomponent copper alloys. *FMM*, **44**, (1), 99–104 (in Russian).
9. Vasil'ev, M. A., Kaminsky, G. G., Pan, V. M. *et al.* (1977) Effect of additions on the superconducting properties of Nb films. *UFZh*, **22**, 1028–1031 (in Russian).
10. Blanchard, B., Hilleret, N., and Quoirin, J. B. (1972) Application of ionic microanalysis to

the determination of boron depth profiles in silicon and silica. *JI. Radioanal. Chem.* **12**, 85.

11. Guthrie, J. W. and Blewer, R. S. (1972) Improved 'tuning' of ion microprobes using scandium thin films targets. *Rev. Sci. Instrum.* **43**, 654–655.

12. Blewer, R. S., and Guthrie, J. W. (1972) Means of obtaining uniform sputtering in an ion microprobe. *Surf. Sci.* **32**, 743–747.

13. Croset, M. (1972) Quantitative analysis of boron profiles in silicon using ion microprobe mass spectrometry. *J. Radioanal. Chem.* **12**, 69–75.

14. Morabito, J. M., and Lewis, R. K. (1973) Secondary ion emission for surface and in-depth analysis of Ta thin films. *Anal. Chem.* **45**, 269–280.

15. Narusawa, T. and Komiya, S. (1974) Composition profile of ion-plated Au film on Cu analysed by AES and SIMS during Xe ion bombardment. *J. Vacuum Sci. Technol.* **11**, 312–316.

16. Liebl, H. (1975) Secondary ion mass spectrometry and its use in depth profiling. *J. Vacuum Sci. Technol.* **12**, 385–391.

17. Morabito, J. M. (1975) A comparison of Auger electron spectroscopy (AES) and secondary ion mass spectrometry (SIMS). NBS Special Publication No. 427, *Secondary ion mass spectrometry*, pp. 33–61.

18. Narusawa, T., Satake, T., and Komiya, S. (1976) Composition of binary alloys by simultaneous SIMS and AES measurements. *J. Vacuum Sci. Technol.* **13**, 514–518.

19. Benninghoven, A. (1976) Characterization of coatings. *Thin Solid Films* **39**, 3–23.

20. Giber, J. (1976) SIMS applications in the investigation of surfaces, thin films and sandwich structures with special regard to quantitative analysis. *Thin Solid Films* **32**, 295–301.

21. Fuller, D., Colligon, J. S., and Williams, J. S. (1976) The application of correlated SIMS and RBS techniques to measurement of ion implanted range profiles. *Surf. Sci.* **54**, 647–658.

22. Wittmaack, K. (1976) high-sensitivity depth profiling of arsenic and phosphorus in silicon by means of SIMS. *Appl. Phys. Lett.* **29**, 552–554.

23. Hofer, W. O., Liebl, H., Roos, G., and Staudenmaier, G. (1976) An electronic aperture for in-depth analysis of solids with an ion microprobe. *Int. J. Mass Spectrom. Ion Phys.* **19**, 327–334.

24. Hofmann, S. (1977) Depth resolution in sputter profiling. *Appl. Phys.*, **13**, 205–207.

25. Wittmaack, K. (1977) Raster scanning depth profiling of layer structure. *Appl. Phys.* **12**, 149–156.

26. Barsony, J., Marton, D., and Giber, J. (1978) Secondary ion mass spectrometry depth profiling and simultaneous electrical investigation of MOS structures. *Thin Solid Films*, **51**, 275–285.

27. Müller, G., Trapp, M., Schimko, R., and Richter, C. E. (1979) Measurement of range distribution of zinc and nitrogen ions in multiple layer substrates with secondary ion microprobe. *Phys. Status Solidi*, **51**, 87–92.

28. Cherepin, V. T. and Vasil'ev, M. A. (1975) *Secondary ion–ion emission from metals and alloys.* Naukova dumka, Kiev, (in Russian).

29. Krimmel, E. F. and Pflederer, H. (1973) Implantation profiles modified by sputtering. *Radiat. Eff.* **19**, 83–85.

30. Storbeck, F. (1978) Zum Einfluss des primärionenstrahlprofils bei der Monolagenanalyse mittels der Methode SIMS. *Krist. Techn.* **13**, 331–341.

31. Kiko, J., Müller, H. W., Büchler K. *et al* 1979] The gas ion probe: a novel instrument for analysing concentration profiles of gases in solids. *Int. J. Mass Spectrom. Ion Phys.* **29**, 87–100.

32. Larikov, L. N., Cherepin, V. T., Vasil'ev, M. A. *et al.* (1974) To lithium vapour effect on the structure of single crystalline refractory metals and alloys. *Fiz. Khim. Mekh. Mater.* **10**, 34–38 (in Russian).

33. Rüdenauer, F. G. (1984) Spatially multidimensional SIMS analysis. *Surf. Interface Anal.* **6**, 132–139.

34. Rüdenauer, F. G. and Steiger, W. (1981) A further step towards three-dimensional elemental analysis of solids. *Mikrochim. Acta.* **2**, 375–389.

35. Steiger, W., Rüdenauer, F., Gnaser, H., Pollinger, P., and Studnicka H. (1983) New developments in spatially multidimensional ion microprobe analysis. *Mikrochim. Acta.* Suppl. 10, 111–117.

36. Morrison, G. H. and Moran, M. G. (1984) Image processing SIMS. In: *Secondary ion mass spectrometry, SIMS IV,* Springer, Berlin, pp. 178–182.

37. Fassett, J. D. and Morrison, G. H. (1978) Digital image processing in ion microscope analysis: study of crystal structure effects in secondary ion mass spectrometry. *Anal. Chem.* **50**, 1861–1866.

38. Suzuki, T. and Tsunoyama, K. (1984) Automation of an ion microprobe mass analyser. In: *Secondary ion mass spectrometry SIMS IV*, Springer, pp. 189–191.
39. Mashiko, Y., Tsutsumi, K., Koyama, H., and Kawazu, S. (1984) Evaluation of metal interaction by color display SIMS technique. In: *Secondary ion mass spectrometry, SIMS IV*, Springer, Berlin, pp. 183–185.
40. Odom, R. W., Fürman, B. K., Evans, C. A., Jr., Bryson, C. E. *et al.* (1983) Quantitative image processing system for ion microscopy based on resistive anode encoder. *Anal. Chem.* **55**, 574–578.
41. Odom, R. W., Wayne, D. H., and Evans, C. A., Jr. (1984) A comparison of camera-based and quantized detectors for image processing on an ion microscope. In: *Secondary ion mass spectrometry, SIMS IV*, Springer, Berlin, pp. 186–188.
42. Rüdenauer, F. G. (1982) Instrumental aspects of spatially 3-dimensional SIMS analysis. In: *Secondary ion mass spectrometry, SIMS III*, Springer, Berlin, pp. 2–21.
43. Wittmaack, K. (1984) Beam-induced broadening effects in sputter depth profiling. *Vacuum*, **34**, 119–137.
44. Brown, J. D., Robinson, W. H., Shepherd, F. R., and Dzoiba, S. (1984) Practical limitations in depth profiling of low energy implants into amorphised and crystalline silicon. In: *Secondary ion mass spectrometry, SIMS IV*, Springer, Berlin, pp. 296–298.
45. Wittmaack, K. (1985) Assessment of the relative contribution of atomic mixing and selective sputtering to beam induced broadening in SIMS depth profiling. *Nucl. Instrum. Meth. Phys. Res.* **B7/8**, 750–754.
46. Patkin, A. J. and Morrison, J. H (1982) Secondary ion mass spectrometric image depth profiling for three-dimensional elemental analysis. *Anal. Chem.* **54**, 2–5.

Chapter 5
Study of processes at the surface

5.1. Adsorption and catalysis

As has been shown in preceding chapters, the secondary ion emission depends strongly on the surface composition. SIMS has therefore been successfully used for the study of various processes that occur at the surface and its interaction with the environment. Such processes include adsorption, catalysis, and oxidation.

Since the main events in the initial stages of these processes occur in the first monolayer, knowledge of the elemental and chemical composition of just this layer is of prime importance. Such methods as Auger electron spectroscopy (AES) or X-ray photo-electron spectroscopy (XPS) provide direct information on elemental composition, but only indirect data on the presence and nature of chemical compounds (from the chemical shifts). SIMS makes it possible to obtain a wide range of data on the presence of different compounds and the dynamics of their interactions with each other and with the surface under consideration.

Analysis of the first monolayers and the study of their chemical state is accomplished by SIMS in the so-called static mode, which is based on the use of a very low primary current density. In this case the surface under study is sputtered very slowly in the course of the ion bombardment. Let θ be a relative coverage of the surface by the first monolayer. Then the variation of θ with time t may be defined from the relation

$$\frac{d\theta(t)}{dt} = \frac{N_0 S}{N} \theta(t) \tag{5.1}$$

where N_0 is the number of primary ions incident on the surface per cm^2 of the surface in unit time; S is the sputtering yield, atom/ion^{-1}; and N is the number of atoms per cm^2 of the monolayer.

Integration gives the time dependence in the form

$$\theta(t) = \exp(-t/T) \tag{5.2}$$

Here $T = N/N_0 S$. During the time T the degree of coverage becomes e times smaller. It follows from this calculation that for a primary ion current density of 10^{-9} A cm^{-2} and a value of S of 10 atom ion^{-1}, the required time T is 10^5s, i.e. the monolayer is preserved for several hours, and hence the mode may be regarded as a static one. Benninghoven [1] was the first to introduce this term. Fogel [2] disapproved of division of SIMS into static and dynamic modes and considered this to be erroneous, since the ion bombardment always causes damage to the object, and so the process is always dynamic. Nevertheless the term is in use, and in the modern literature SIMS of the surface at the monolayer level is, as a rule, referred to as static and sometimes is specially designated as SSIMS (static secondary ion mass spectroscopy).

Fulfilment of the condition (5.2) indicates the presence of a monolayer coverage. If plotting of the natural logarithm for the secondary ion current of the adsorbate as a function of time gives the straight line it implies that the coverage does not exceed the monolayer [3].

The first attempts to apply SIMS for study of adsorption, catalysis, and oxidation processes were made in the sixties in the Fogel laboratory [4–10]. Several examples are given below of SIMS analytical possibilities for study of adsorption and catalysis processes at the surface.

5.1.1. Oxygen Adsorption on Molybdenum Surface

These processes were analyzed by SIMS in the course of Mo target bombardment by Ar$^+$ ions with an energy of 4 keV and a current density of " 10^{-7} A cm^{-2} (static mode) [11]. Oxygen partial pressure in the residual gas was varied from 4×10^{-6} to 10^{-4} Pa at Mo ribbon temperatures of 300–1900 K. At 2000 K the secondary ion mass spectrum contains no ions other than Mo$^+$, which is evidence of target surface purity. The temperature was then decreased to a preselected level, and the isoterms of the secondary ion current $I_i^+(t)$ were studied for O$^+$, O$_2^+$ and molybdenum oxide ions. The moment when the Mo ribbon reaches the given temperature was taken as zero time for plotting $I_i^+(t)$ curves. It took less than 20 s to cool the ribbon from 2000 to 500 K. When the residual gas oxygen was adsorbed on the ribbon, whose surface temperature was not below 1300 K, the secondary emission mass spectrum contained only O$^+$ ions in addition to Mo$^+$ ions ejected from the target. It thus may be assumed that oxygen adsorption on a molybdenum surface is accompanied by the dissociation of adsorbed oxygen molecules into atoms. The $I_i^+(t)$ curve for O$^+$ ions reflects the kinetics of oxygen atom accumulation on Mo surface. The process became saturated in approximately 6 min. Then the adsorption equilibrium became established, and the coverage no longer changed.

The $I_i^+(t)$ curves obtained for different oxygen pressures at high target temperature differ in the time interval needed to reach saturation. This interval is shorter, the higher the oxygen pressure. When oxygen interacts with the Mo surface at lower temperatures the emission of O$_2^+$ ions and a number of Mo oxide ions is observed along with the emission of O$^+$ ions. Isotherms $I_i^+(t)$ are similar for all the molybdenum oxides and start at the same moment, which is much later than the moment of atomic oxygen ion emission. O$_2^+$ ion emission begins simultaneously with the emission of oxide ions. This means that the processes of oxygen adsorption on the Mo surface and of Mo oxidation do not begin at the same time. There is a time interval (latent period) when accumulation of chemisorbed atomic oxygens is the only process on the Mo surface. With further oxygen admission and its interaction with the chemisorbed atomic oxygen the oxides are formed on Mo surface. The moment when saturation is reached in the isotherms corresponds to the formation of a two-dimensional oxide film. Further modifications in the compositions and structure of this film correspond to the development of oxidation processes and will be discussed below.

Another typical approach to the study of initial stages in adsorption and oxidation is as follows. The sample was carefully cleaned by ion etching and heating so that the surface was atomically clean. Then oxygen under a fixed partial pressure (e.g. 10^{-5} Pa) is admitted into the chamber, and the exposure of a definite duration is made. The exposure is measured in Lengmuirs (1 L = 1.33×10^{-4} Pa·s). Spectra are then taken in static SIMS mode and are interpreted for identification of characteristic peaks of relevant oxides. Figure 5.1 (from [12]) shows dependence of ion currents for different Mo oxides on the exposure of pure Mo surface to oxygen, as obtained by Benninghoven. It may be seen that all the ion currents reach saturation whose level is used for curve normalization. Currents of ions with lower oxygen content, e.g. Mo O_2^+, formed from lower oxides, became saturated at smaller oxygen exposures than ions with larger oxygen content, e.g. Mo O_3^+. It has also been found that the current magnitude depends on the dose but, in the static SIMS, does not depend on time.

After an exposure of approximately 100 L all the secondary ion currents cease to depend on the exposure: the saturated layer is perhaps formed distinguished by a definite coverage degree and stoichiometry. These results for Mo and many other metals [13–16] lead to the conclusion that oxygen

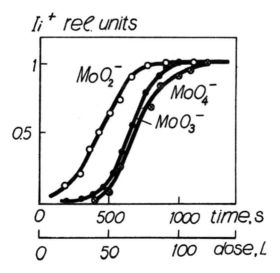

Figure 5.1 Variation of ion currents of different Mo oxides with dose for a clean Mo surface exposed to oxygen at 10^{-5}Pa.

admission leads to formation of 'metal-oxide' layers with different structure. Each structure gives secondary molecular ions possessing a definite intensity ratio displayed in the characteristic spectra [17] ('fingerprint spectra'). Thus if chromium is exposed to 50–100 L of oxygen, a monolayer of chromium oxide is formed characterized by the emission of CrO_2^- ions. Cr atoms in this ion have a weak positive valency. Larger exposures result in another oxide structure characterized by the emission of CrO^+ ions. The probability of positive secondary ions being emitted from this compound increases with the growth of Cr atom positive valency. The latter manifests itself in chemical shifts in Auger electron spectra as in the case of Ta and V. Contrary to earlier works by Fogel *et al.* [15, 18], measurements are now made in cleaner conditions although the idea of the experiment remains unchanged.

5.1.2. Catalytic Reactions

Applications of SIMS to the study of heterogenic catalytic reactions on solid surfaces seem to be rather promising [4, 10]. The technique of such investigations was first developed by Fogel *et al.* and is based on simultaneous measurement of secondary ion emission and mass spectrometric analysis of gases surrounding the bombarded sample [19–23].

As an example of SIMS application to the study of catalytic reactions the results obtained from investigation of ammonium dissociation and synthesis on iron [22, 23] will be considered.

The catalyst on which these processes were studied had the shape of a ribbon and was made of pure iron (99.99%). The course of the reaction was monitored through observation of NH_3^+, H_2^+ and N_2^+ ion production from ammonium ionization by the electron impact, depending on the catalyst temperature. It has been established that the reaction of ammonium dissociation does not occur on the surface of iron catalyst which had not been treated beforehand.

Consideration of mass spectra of secondary ions ejected from the untreated iron catalyst surface has revealed large number of peaks originating from the presence of oxygen and iron oxides at the surface. The ions $^{16}O^+$, $^{32}O_2^+$, $^{56}Fe^+$, $^{72}FeO^+$, $^{88}FeO_2^+$, $^{112}Fe_2^+$, $^{128}Fe_2^+$, $^{144}Fe_2O_2^+$, $^{160}Fe_2O_2^+$, $^{176}Fe_2O_4^+$, $^{200}Fe_3O_2^+$, $^{216}Fe_3O_3$, $^{232}Fe_3O_4^+$ were observed. They may be found in different ratios in the temperature range 20–800°C. Just the presence of oxides is believed to make the iron catalyst inactive to the ammonium dissociation reaction. This assumption can be checked by reduction of the surface. With this aim the iron ribbon was heated for 4 h in hydrogen under the pressure of several hundreds Pa. After such treatment the iron oxides have not been discovered at temperatures above 500°C. The iron catalyst became distinctly active to ammonium dissociation reaction in the atmosphere where hydrogen partial pressure was at the level of 10^{-2} Pa.

The same Fe catalyst was used to study ammonium synthesis by SIMS. It has been found that iron which catalyzes ammonium dissociation is equally active in the synthesis process. This follows from the isobars $I_i^*(t)$ for NH_3^+ ions. Already at 200°C the compound NH_3 appears on iron surface. With further rise in the catalyst temperature the coverage of its surface by NH_3 molecules attains the maximum near approximately 400°C, and then is decreased monotonously.

Elementary processes that occur at the catalyst surface can be understood from the analysis of NH_3^+, NH_2^+, and NH^+ ion yields (these ions characterize the NH_3 molecule and its fragments) and H_2^+, H^+, N_2^+, N^+ ion yields (these ions are characteristic to the molecules of reactive gases and their fragments) as a function of temperature. As has been demonstrated in the course of experiment, at the ammonium synthesis reaction with the catalyst surface FeN_2^+ ions are ejected. The temperature dependence of these ions was plotted as also the dependence of NH_3^+ and FeN_3^+ emission intensity on nitrogen and oxygen pressure.

All these dependences are given in Fig. 5.2. It may be concluded from the consideration of the temperature dependence of N_2^+, N, H_2^+, and H ion emission that N_2 and H_2 molecules do not dissociate into atoms. The shape of the curve for N_2^+ ions indicates that these ions are ejected from both the adsorbed N_2 molecules and from the FeN_2 molecules. The intensity of N_2^+ ion emission is increased at temperatures above 600°C because of higher coverage of the catalyst surface by N_2 molecules created in the ammonium dissociation reaction. The fact that $I_i^+(t)$ curves are similar for NH_3^+ and FeN_2^+ ions allows us to suggest that FeN_2 formation is the first stage of the ammonium synthesis reaction. The conclusion about the second stage of this reaction may be made from the comparison of $I_i^+(t)$ curves for NH_3^+, NH_2^+, and NH^+ ions. The ratio $I_{NH_2}^+/I_{NH_3}^+$ does not depend on the catalyst temperature in the range 20–800°C. Hence, NH_2^+ ion may be regarded as the fragment of NH_2 molecule. However, the ratio $I_{NH}^+/I^+_{NH_3}$ varies with the catalyst temperature through

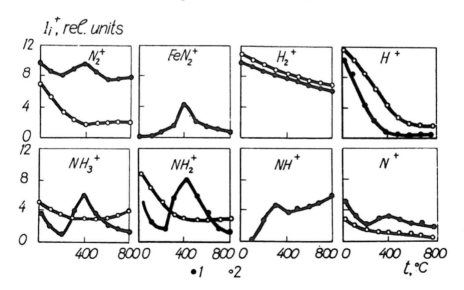

Figure 5.2 Temperature dependence $I_i^+(t)$ for secondary ions of active (1) and inactive (2) catalysts in a mixture of nitrogen and hydrogen at $P_{N_2} = 2 \times 10^{-3}$Pa and $P_{H_2} = 6 \times 10^{-3}$Pa.

the whole range 20–800°C. This might imply that NH_3 molecules are not the only source of NH^+ ions. Another conceivable source can be an NH radical produced in the reaction

$$FeN_2 + H_2 \rightarrow Fe \underset{\diagdown}{\overset{\diagup}{}} \begin{array}{c} NH \\ \\ NH \end{array}$$

In these terms the shape of $I_i^+ (t)$ curve for NH^+ ions may be explained as follows. At the temperatures 100–400°C the Fe–NH complexes and NH_3 molecules on the catalyst surface determine the curve run. In this temperature range the NH^+ ions are ejected only from these particles. Consideration of the curves in Fig. 5.2 leads to the conclusion that the third stage of ammonium synthesis proceeds according to the reaction

$$Fe - NH + H_2 \rightarrow Fe - NH_3$$

Some assumptions can be made from the pressure dependence of the curve; the catalyst coverage by FeN_2 molecules tends to saturation as the pressure increases. The same has been observed for NH_3 coverage. The ammonium yield is linearly increased with the rise in hydrogen pressure. Thus the mechanism of ammonium synthesis on iron has been elucidated using SIMS data.

A similar approach has since been used by various authors to study other catalytic reactions and to solve many theoretical and practical catalysis problems, as well as the problems of technology monitoring [24, 25].

5.2. Investigation of metal oxidation

Corrosion processes, in particular metal oxidation, differ essentially from catalitic ones: they occur first in the outer monolayer and then, by means of diffusion of participating particles, penetrate into the metal bulk so that an oxide layer of significant thickness is formed.

Metal oxidation processes have been the subject of numerous works [24]. However, they have been studied, as a rule, at the stage where the oxide film thickness on the metal surface exceeded several tens of monolayers, since the methods used for corrosion investigation were inadequate for the study of the initial oxidation stage, i.e. the stage where the true surface chemical compounds were formed.

Modern SIMS-based methods for oxidation and corrosion study have been developed by Fogel [4, 10] and successfully realized in his laboratory [5–8, 17, 26–29]. According to his method the mass spectra of secondary ions ejected from the surface under study at different target temperatures are measured as a function of the nature and pressure of the gas media in which target is placed.

Consider as an example the results of the Mo oxidation study [17], which are interesting from the methodological point of view.

Ar^+ ions with an energy of 4 keV and a current density of $5 \times 10^{-7} A \, cm^{-2}$ were used for the bombardment. Secondary ion mass spectra taken of the Mo target at room temperature revealed positive and negative ions of Mo and its oxides Mo^+, MoO^+, MoO^-, MoO_2^+, MoO_2^-, Mo_2^+, Mo_2O^+, $Mo_2O_2^+$, $Mo_2O_3^+$, $Mo_2O_4^+$, $Mo_2O_5^+$, and $Mo_2O_6^+$. It has been established that the emission intensity of these ions varies with changes in Mo ribbon temperature and oxygen pressure. Figure 5.3 shows $I_i^+(t)$ for all the Mo oxide ions produced at the interaction between residual oxygen gas and the Mo surface. The target surface was beforehand cleaned of adsorbed particles and oxides by heating to 1900 K. At this temperature only Mo^+ secondary ions were ejected from the surface of the pure metal. The temperature was then lowered to a certain level, and the current magnitude I_i^+ corresponding to the steady state of the oxide film on the Mo surface (found from the kinetic curves $I_i^+(t)$ where t is the oxygen adsorption duration) was used to plot these dependences.

One cannot judge the oxide film composition on the Mo surface directly from the mass spectrum, since the latter contains not only ions of Mo oxide molecules but also fragment ions created at the dissociation of the parent molecule after irradiation by primary ions. However, the oxide film composition can be defined accurately enough from consideration of $I_i^+(t)$ for different ion species.

In static SIMS the number of molecules of any oxide, n_i, is practically independent on the bombardment duration, but depends only on temperature and pressure. Since the ionization coefficient α_i has been shown to be temperature independent, the secondary ion current I_i^+ is a simple function of n_i and α_i:

$$I_i^+ = \alpha_i n_i(p,t) \qquad (5.3)$$

while the secondary ion total current I^+ is a linear superposition of all the ion currents:

$$I^+ = \Sigma \, \alpha_i n_i(p,t) \qquad (5.4)$$

If two kinds of secondary ion originate from a molecule of the same surface

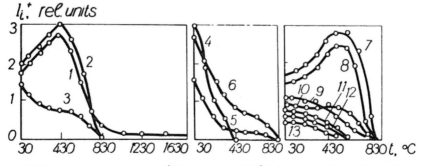

Figure 5.3 Temperature dependence $I_i^+(t)$ at $P_{O_2} = 3 \times 10^{-6}$ Pa for various Mo and Mo oxide secondary ions. (1) Mo^+; (2) MoO^+; (3) MoO_2^+; (4) MoO_2^-; (5) MoO_5^-; (6) MoO_2^-; (7) Mo_2^+; (8) Mo_2O^+; (9) $Mo_2O_2^+$; (10) $Mo_2O_3^+$; (11) $Mo_2O_5^+$; (12) $Mo_2O_6^+$.

compound having concentration $n_i(p,t)$, the current ratio for these ions should not depend on temperature, since

$$I_1^+(t) = \alpha_1 n(p,t), \ I_2^+(t) = \alpha_2 n(p,t) \ \text{and}$$
$$I_1^+(t)/I_2^+(t) = \alpha_1/\alpha_2 = \text{const.} \tag{5.5}$$

In this case the curves of temperature dependence should have the same run for ions originating from the same surface compound. This fact was used in the analysis of oxide compositions: the analysis should commence with the heaviest ions discovered in the spectrum. Since the $I_i^+(t)$ curve is similar for lighter ions it may be assumed that these ions are fragments of larger molecules and are not worth considering. Thus it follows from the curves in Fig. 5.3 that the oxides Mo_2O_6, MoO_3, MoO_2, Mo_2O_3, and Mo_2O are present on the Mo surface.

The above method was later used to study Mo oxidation at high oxygen pressures and to analyse the interaction between oxygen and W, Nb, Ag, Fe, Be, GaAs, and other metals, alloys, semiconductors and their compounds.

Observation of the $I_i^+(t)$ curve run as a function of temperature serves as a basis for subdivision of ions of different structures into spectra characteristic for a given upper compound. An alternative method is to compare SIMS spectra obtained from surface oxides and from relevant samples of monolitic oxides of a known composition.

Investigation of different Cr oxides in the form of bulk speciments or thin films has shown [12] that:

(1) The mass spectra of different Cr oxides contain the same fragments. A specific chemical compound is characterized by a typical abundance of these fragment ions, reflecting the environment of the atom in the lattice, the nature and number of nearest neighbours, the bond strength, the dissociation energy, electronic configuration, etc.

(2) For a primary ion energy of several keV the concentration ratio of Cr and O for fragment ions does not immediately reflect the stoichiometry of the original compound. But when primary ions with an energy of several hundred eV are used, a direct correlation between the fragment ion composition and the arrangement of atoms in the lattice is observed [28].

(3) The principle of superposition of characteristic spectra, applied in the gas mass spectrometry, can also be applied in SIMS.

The above regularities help to essentially simplify the interpretation of usually complex SIMS spectra. For example, Fig. 5.4 illustrates the superposition of characteristic SIMS spectra from a Cr_2O_3 bulk sample (solid lines) and from oxidized Cr (dotted line) [12]. Relative intensities of fragment and other ion emissions are seen to coincide well enough, suggesting the presence of oxides of a given type in the spectrum from an oxidized compound of unknown composition. The use of this method is also illustrated in Figs 5.5 and 5.6 [12]. As may be seen from Fig. 5.5, the mass spectrum from the oxidized Cr surface is difficult to analyse. In the course of bombardment some peaks are increased by several orders of magnitude, while the others are decreased to the same extent. In this case any conclusion about the depth variation of stoichiometric composition is practically impossible. However, if the layer under analysis is assumed to be composed of three different Cr oxides, each having its own fingerprint spectrum, then this bundle of lines can be easily unravelled. Expansion of the whole spectrum into characteristic wavelengths has revealed that the upper layer contains pure Cr and two oxides, CrO and Cr_2O_3, which are distributed through the oxidized layer thickness (see Fig. 5.6).

It should be noted that SIMS was not immediately acknowledged as a reliable method for the study of adsorption and oxidation phenomena. Many authors preferred traditional flash methods (thermo-induced desorption) or more customary techniques of photoelectron and Auger electron spectroscopy. Use of SIMS in combination with other methods, which may be regarded as reference methods, helped greatly in establishing SIMS and estimating its potential. Much work has recently been done in this area [31–38]. However, the largest contribution was made by the group under Benninghoven. Various stages of oxidation of V, Ni, Ti, Mo, Co, Mg, Sr, Ba, Fe, and Al have been studied, and it has been definitely concluded that SIMS yields rich information on the mechanism of oxygen interaction with metals, the oxygen state on the surface and in the bulk, as also about the dynamics of various phases [35–37]. SIMS has also been found to give quantitative information about oxygen concentration at the surface (the coverage degrees in a monolayer range). Figure 5.7 compares different signals characterizing the oxygen concentration at the surface of polycrystalline Ti after different doses of oxygen exposure (taken from [34]). Signals of O^- ions in SIMS (3), O KLL in AES (1) and O ls in XPS (2) are normalized to their maximum values and are dose-identical functions within the limits of measurement accuracy. Combined investigation of oxidation with the use of modern instrumentation will be illustrated by the example of the Ti oxidation study carried out by Benninghoven *et al.* [34]. Pure Ti foil (0.1 mm thick) and 99.6% Ti was used

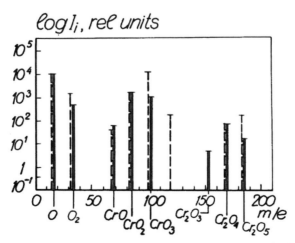

Figure 5.4 Characteristic spectra of negative secondary ions of bulk oxide Cr_2O_3 (solid lines); and oxidized Cr surface (dashed lines).

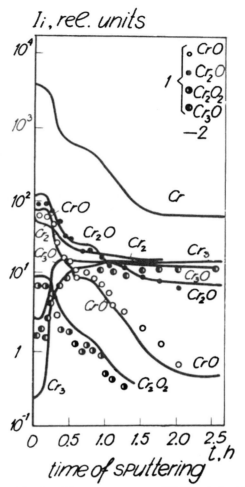

Figure 5.5 In-depth modification of the mass spectrum at SIMS analysis of the oxidized layer on Cr surface. (1) Calculated data; (2) Measurement.

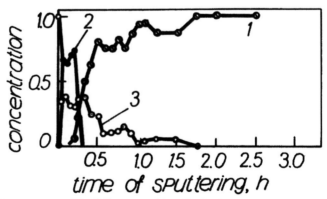

Figure 5.6 Mass spectrum (Fig. 5.5) converted into distribution profiles of various Cr oxides according to their finger print spectra (1) Cr; (2) CrO; (3) Cr_2O_3.

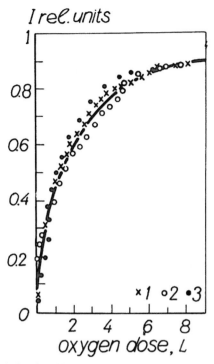

Figure 5.7 Normalized relative signal intensity in (1) AES, (2) XPS, and (3) SIMS during oxygen concentration analysis at oxygen-exposed clean Ti surface.

as a target. It was cleaned by ion etching with a total dose of 10^{-1}–10^{-2} A s cm^{-2}. The surface was then oxidized by the exposure to an oxygen atmosphere under a pressure of 10^{-6}Pa. Oxidized layer was etched away, where necessary, for depth analysis by means of an ion beam with a current density of 10^{-7}–10^{-5} A cm^{-2}. Surface composition was controlled by SIMS at 3 keV Ar^+ ion bombardment in the static mode and also by the use of CMA Auger electron spectrometer provided with Al K_α source. The base pressure in the system was 5×10^{-9}Pa. It rose to 5×10^{-7}Pa when the SIMS ion gun was in operation. Different types of measurement were performed over a 1 s interval, that is quasi-simultaneously. The effects of different lengths of oxygen exposure on the results obtained by SIMS and AES are compared in Fig. 5.8. The change in the relative emission intensity of secondary ions of the $TiO_n^{+/-}$ type is shown in Fig. 5.8(a). The characteristic succession in the appearance of certain molecular ions corresponds, the authors of [34] believe, to the increase in the mean valency of Ti atoms in the surface layer, so that the maxima in the emission of TiO_n^+ and TiO_n^- molecular ions are shifted to larger n with longer oxygen exposure. Therefore the oxygen-enriched ions of the TiO_3^- and TiO_2^+ type appear last, and the emission intensity of TiO ions becomes lower after reaching a maximum. Characteristic modifications in secondary ion spectra are terminated when a dose of the order of 20 L is reached. The succession of appearance and disappearance of characteristic

SIMS of solid surfaces

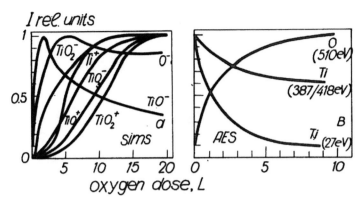

Figure 5.8. Relative intensity of characteristic secondary ions of different oxides (a) and Auger-electrons (b) as a function of oxygen exposure during oxidation.

ions in the mass spectrum is inverted for the depth analysis of the produced layer.

AES measurements were carried out by means of observation of the O_2 KLL transition at 510 eV, and some other transitions in Ti ($M_{2.3}VV$ at 27 eV, $L_{2.3}M_{2.3}M_{2.3}$ at 387 eV and $L_{2.3}M_{2.3}V$ at 418 eV). Figure 5.8(b) gives relative peak–peak amplitudes in Auger spectra of these transitions as a function of the dose. O KLL signal grows fast and attains saturation at the dose about 10 L. Transition at 27 eV in Ti disappears almost completely, but the signals at 387 and 418 eV preserve almost 60 % of their intial value. Essential decrease in the signal at 27 eV is due, perhaps, to the fact that only in this transition the valence band participates twice, but just this band suffers electron deficiency from the very beginning of the oxidation.

XPS investigations involve the emission intensity measurement from Ti $2p$ core levels. The signal was found to diminish with the exposure growth; the line became slightly broader but no chemical shifts were observed for doses up to 20 l. At 1000 l the intensity of the shifted $2p$ peaks is still comparable withu that of remaining substrate peaks.

Hence, SIMS yields much broader information on both kinetics and mechanism of initial oxidation stages. Chemical information obtained from the measurement of chemical shift in XPS may be obtained only after a three-dimensional structure is built in the oxides several monolayers thick. The absence of chemical shifts in the case of even a monolayer coverage suggests that characteristic alterations in electronic structure of core levels resulting in chemical shifts can occur only if metal atom is actually within the chemically active environment.

SIMS can also be helpful in the study of surface diffusion especially at low concentrations of diffusing element. The technique of such measurements has been developed under the guidance of Fogel [39] and is based on observation of the $I_i^+ (\tau)$ curve (τ is the diffusion duration) for the secondary ions corresponding to atoms which diffuse over the substrate surface.

5.3. Comparison of SIMS with other methods of surface analysis

Like any other method of surface analysis, SIMS provides only limited information on composition and structure of the surface and therefore profits from combination with complementary techniques [40]. Many of these techniques have been developed to a high degree of perfection and provide various information on composition, crystal and electronic structure, chemical bonding of the surface, etc. This is especially true for various kinds of electron spectroscopy, low and high-energy electron diffraction, and different ion spectroscopies. Instruments used to realize these methods are commercially available and are currently applied not only in scientific investigations but also for the control and development of modern technological processes. These methods and instruments are described in detail in [40–49].

Combinations of various methods with SIMS may be aimed at two objectives. In the first case physical aspects of secondary ion formation are the focus of interest, and hence SIMS should be combined with the analysis of optical radiation (SCANIIR), ion neutralization spectroscopy (INS), electron energy loss spectroscopy (LEELS), and similar techniques providing physical information [50]. In the second case, which is of more practical interest, a combination is selected to overcome specific analytical limitations inherent to both SIMS and other methods. For instance, the quantification problems of SIMS and the poor detection limits of methods such as electron spectroscopy have been the most frequent reasons.

The actual analytical performance of a specific combination depends on its instrumental implementation, and reliable results can only be obtained if complementary methods are applied to the same surface simultaneously or sequentially in one UHV system.

Among the many different surface analytical techniques only a few have obtained practical application for combination with SIMS. They are AES, XPS, UPS, TDMS, LEELS, LEED, and ISS [51]. Naturally, each additional method should provide complementary information. Thus, the relative ease of quantitative evaluation of AES and XPS provide such additional features to SIMS, which we have seen has sensitivity and a wealth of signals related to surface chemistry, but poor accuracy. Evaluation of chemical shifts in electron spectroscopy can be used to clarify the molecular SIE from inorganic compound surfaces. Combinations of methods which give information from different depths, such as ISS, SIMS, AES, and XPS, can be useful for investigation of near-surface concentration distributions e.g., for out-diffusion, segregation, or preferential sputtering studies.

Information depth is a very important parameter for comparison and combination of complementary methods. As has been shown in Chapter 4, the depth resolution of SIMS is limited (from the physical point of view) by the volume within which the collision cascade is developed and by the depth from which secondary ion emission is possible (about 1 nm). For static SIMS the main physical mechanism is perhaps direct momentum transfer to a surface atom or molecule. This allows one to obtain information about the first surface layer. The same is valid for the case of ion scattering spectroscopy.

For most kinds of electron spectroscopy the characteristic information on the surface under study is contained in the magnitude of electron energy and momentum. Therefore information may be lost due to elastic and inelastic scattering of electrons on their way from the site of their ejection towards the surface. Inelastic scattering processes result from electron interaction with other electrons or phonons. However the energy loss due to electron–phonon interaction is too small to be observed with the energy resolution attained in the experiment. For dense materials the inelastic scattering probability is proportional to the path length x in a solid. Therefore the flow of electrons I_0 possessing a definite energy and momentum decreases exponentially as:

$$I = I_0 \exp\left(-x/\lambda\right) \tag{5.6}$$

where λ is the electron mean free path length.

All the available experimental data on the mean free path for various materials (metals, predominantly) are summarized in Fig. 5.9 [49]. These data were obtained either by measurement of the moment when characteristic electrons were emitted from different substrates on which the layers of another material had been deposited or, in some cases, by direct transmission measurements.

The electron mean free path is not necessarily equivalent to depth resolution. In appearance potential spectroscopy and in characteristic energy-loss spectroscopy the primary electrons themselves serve as the signal-forming electrons. They penetrate the surface layer twice, so that the surface locality depth resolution equals half the free path length. For electrons ejected from the surface at some angle φ to the normal, the information depth is proportional to $\lambda \cdot \cos \varphi$.

In characteristic energy-loss spectroscopy the information depth in the case of surface excitations has, in general, no connection with the mean free path length but is determined by the electron energy, reflection angle, and

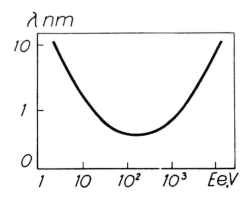

Figure 5.9 Averaged mean free path length of electrons in different materials.

spectrometer angular resolution. It may hence be concluded that the information depth in electron spectroscopy is a rather uncertain quantity. This should be kept in mind when making quantitative measurements. It may be stated, in general, that the information depth is less defined the higher the surface sensitivity. Therefore higher-energy electrons are preferable for quantitative measurements, at the cost of surface sensitivity.

It is a common belief that AES is more advantageous than SIMS because of the non-destructive nature of the electron interaction with the sample surface, and because the electron beam does not contaminate (and thus does not modify) the surface composition, this being inevitable in the case of ion bombardment. But this opinion is not supported by practice. It turns out that in order to get AES of sufficient sensitivity, the density of electron beams must be much higher than the ion beam density used in SIMS. But such electron bombardment leads to a rapid modification of the surface due to dissociation of adsorbed molecules, reduction of surface oxides, diffusion to the irradiated area or from it, the electron-induced desorbtion [33]. This effect of electron bombardment limits the electron irradiation dose to about 2×10^{-4} A s cm^{-2}. Only below this value is AES truly a non-destructive (static) method. But this limit is about 10^5 below the doses generally used in standard or imaging AES. The situation can be improved by using direct secondary electron counting at a primary electron current density of 10^{-8}–10^{-7} A cm^{-2}. In this case the above-stated limit for the irradiation dose is not exceeded even at a bombardment duration of up to 10^3 s. However even in this case the sample may be contaminated by negative ions contained in the primary electron beam. The ions are formed from O, F and Cl atoms sputtered or desorbed from ceramic parts and filaments of the electron gun [52]. These atoms are ionized by electron impact, accelerated, and implanted into sample surface. Here they may be detected by SIMS and removed by ion sputtering, thus providing one more good reason for combining electron and ion probing methods.

An important requirement for combined analysis is compatibility, in the sense that changes of surface properties due to individual analysis probing must be excluded or controlled. Energy deposition rates and surface damage mechanisms associated with most frequently used complementary methods are shown in Table 5.1 [51]. As can be seen from this table, XPS is always operated in a static mode, whereas SIMS, ISS, and AES have to be implemented in a way that allows static work of sufficient sensitivity, particularly when dealing with adsorbates and surface reactions. From this point of view combined SIMS and AES (in imaging mode) are good for depth profiling but can hardly be used for surface characterization.

For calibration purposes all applied techniques should have a common range of data (e.g. on composition) and no special surface properties should be required for their application. As SIMS in most of its typical applications destroys the surface crystal order, this limits application of LEED, UPS, ELS and similar methods, which show their full potential only on single crystalline surfaces.

Table 5.1
Energy flux per lattice site of excited volume and surface damage mechanism for some analytical techniques

Technique	$eV\,s^{-1}$	Damage mechanism
AES standard	10^5	Desorption by electronic transitions
imaging	10^6	Deposition of contaminants
static	10^2	Bombardment induced reactions, heating
SIMS standard	10^5	Sputtering, bombardment induced reactions,
imaging	10^6	Atomic mixing,
static	10^{-5}	Heating
ISS	as SIMS	As SIMS
UPS	10^{-1}	Deposition of contaminants
XPS	10^{-2}	As UPS
LEED	10^5	As AES
ELS	10^{-5}	As AES
TDMS	10^{-1}	Deposition, in-diffusion of contaminants

From the above criteria the combination of SIMS with AES, XPS, ISS, and TDMS seems to be most useful. Comparison of the main analytical properties of these methods is given in Table 5.2.

One important feature concerns the type of species analysed. Whereas AES, XPS, and ISS detect species at the surface, SIMS and TDMS analyze species *removed* from the surface. Thus, in depth profiling with an effect of preferential sputtering, AES, XPS, and ISS give correct data at the beginning only, while SIMS at this point is wrong. In sputtering equilibrium SIMS measures particles reflecting sample stoichiometry, whereas the other methods are affected by variation in the composition of the altered layer created by preferential sputtering. Improved quantification techniques are, however, required in order to exploit this advantage of SIMS.

From the technical point of view, the most important prerequisite for successful operation of combined method instruments is convenience of operation. For instance, operations and the time necessary to switch over from one method to another must be reduced to a level that can be managed within a few seconds in a fail-safe way. This requires computer control of all instrument parameters, in particular the vacuum system, excitation sources, mode of operation of spectrometers and detection systems, sweep parameters, and sample conditioning. Nowadays, completely automatic control of detection systems and sweep parameters is available commercially.

Complete computer control requires an automatic protocol of experimental parameters like excitation currents, pressures, voltages applied to particle optics, etc., so that the analyst can feed experimental results back into further analytical procedures. An advanced approach to this problem is shown schematically in Fig. 5.10 [51]. It relies on specially designed electronic equipment, e.g. excitation sources and secondary ion optic supplies interfaced to an IEEE–488 bus, and a modular vacuum system design.

A schematic diagram of a typical simple apparatus designed for analysis by

Table 5.2
Main analytical characteristics of some surface analysis techniques

Characteristic	Method				
	SIMS	AES	XPS	ISS	TDMS
Static mode	possible	possible	always	possible	no
Element detection	all	$Z \geqslant 3$	$Z \geqslant 3$	$Z \geqslant 3$	H, C, N, O
Isotope detection	all	no	no	possib.	possible
Compound detection	partial	some	some	no	some
Binding state information	partial	some	yes	no	some
Structural information	possible	no	possible	possible	no
Sensitivity, monolayers	10^{-6}	10^{-3}	10^{-2}	10^{-2}	10^{-2}
Dynamic range	10^{8}	10^{3}	10^{2}	10^{2}	10^{1}
Quantification	possible	yes	yes	(yes)	(yes)
Information depth monolayers	$\leqslant 3$	3–8	5–15	1	1

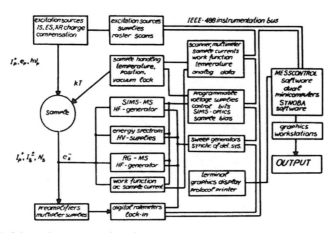

Figure 5.10 Schematic representation of computer control system for a combined method.

combined methods is shown in Fig. 5.11. A spherical 80 l chamber is pumped by cryosorption forepumps of 400 l s^{-1} capacity, a getter ion pump, and a titanium sublimator with a liquid-nitrogen-cooled cryopanel, the whole system being bakable to 520 K. The ultimate vacuum is about 5×10^{-9} Pa. The ion gun is differentially pumped so that a vacuum of not worse than 10^{-7} Pa is kept when the gun is operating. In the static SIMS mode Ar$^+$ ions are used with $E_p = 1.3$ keV and $I_p = 450$ pA. In the dynamic SIMS mode (depth profiling) $E_p = 1.6$ keV and $I_p = 6.5$ nA. The beam spot on the target surface is 3×3.5 mm^2. Secondary ions are analysed by a quadrupole mass filter in the mass range 1–300 a.m.u., with an ionization chamber placed at the entrance to analyse the residual gas composition. In another cross-section of the chamber, the four-grid LEED optics are positioned and used for AES. In addition there is an orifice in the chamber for introduction of an ion or atom beam for use in ion or atom beam scattering spectroscopy [53].

A schematic diagram of a more advanced and sophisticated combined

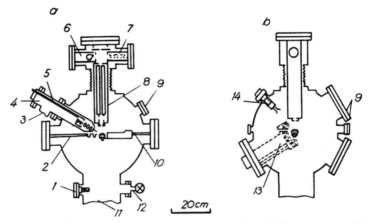

Figure 5.11 Combined SIMS–AES instrument for surface studies. (1) Vacuum gauge; (2) manipulator with Faraday cup; (3) differential pumping of ion gun; (4) ion gun; (5) inlet of working gas; (6) CEM; (7) SEM; (8) quadrupole; (9) viewing port; (10) sample holder; (11) pumping orifice; (12) leak valve; (13) LEED–AES optics; (14) electron gun.

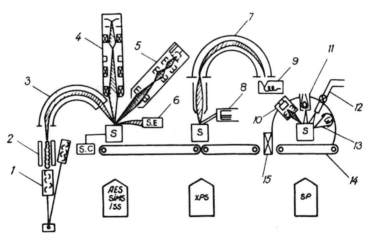

Figure 5.12 Modern combined method instrument for SIMS, AES, XPS, and ISS: the ASIX-1000. (1) SEM; (2) quadrupole; (3) energy analyser for AES, ISS, and SIMS; (4) electron probe; (5) ion probe; (6) secondary electron collector; (7) energy analyser for XPS; (8) X-ray tube; (9) CEM; (10) sputtering ion gun; (11) electron gun; (12) gas inlet; (13) heater; (14) transport system; (15) separating valve. S — sample; SP — sample preparation.

method instrument developed by SHIMADZU ASIX 1000 is shown in Fig. 5.12. The instrument consists of three chambers where the sample is prepared, and analysed by XPS and SIMS, AES and ISS [54].

New trends in surface study techniques and the development of relevant instruments arise from the prospects connected with the investigation of various emission processes with angular resolution. In this case angular-resolved SIMS may be combined with angular-resolved AES and XPS, thus providing new information on the composition, structures, and electronic structure of clean and adsorbate-covered surfaces [55]. A schematic diagram of a combined instrument for these studies is shown in Fig. 5.13.

As a final remark it should be mentioned that sample preparation is a stage of primary importance for all studies of solid surfaces. Subsurface layer composition depends strongly on preliminary treatment in the ambient atmosphere (cutting, polishing, etching, washing, etc.). That is why a rather tedious and complex procedure of surface cleaning and preparation is necessary after the sample has been introduced into the vacuum chamber. The ideal surface must be free from any defects and contaminations.

Most contaminations can be removed by ion etching of the surface. But in this case the bombarding gas is implanted into the sample, and radiation defects are created which must be removed by the sample annealing in UHV. Hence, the surface composition is determined by the net effect of sputtering, evaporation, and diffusion during multiple cycles of etching and annealing. All these processes are especially pronounced when multi-component samples are dealt with [16]. Special difficulties arise most often due to surface

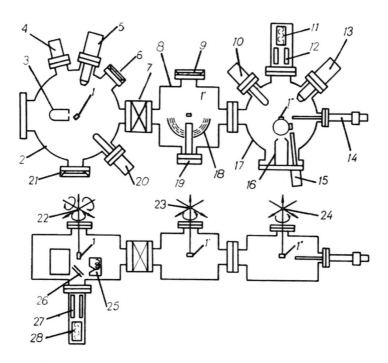

Figure 5.13 Schematic diagram of ion and electron spectrometers with angular resolution. (1) Sample; (2) analytical chamber; (3) energy analyser; (4) electron gun; (5) ion gun; (6) viewing port; (7) valve; (8) LEED chamber; (9) viewing port; (10) electron gun; (11) SEM; (12) quadrupole; (13) ion gun; (14) transport rod; (15) air-lock system; (16) CMA; (17) preparation chamber; (18) LEED optics; (19) LEED electron gun; (20) electron gun; (21) viewing point; (22, 23, 24) sample manipulators; (25) CEM; (26) parallel plate capacitor; (27) quadrupole; (28) SEM.

contamination by carbon and sulphur, for which the surface serves as a kind of getter. Sulphur penetration of a surface is accelerated at high temperatures. Therefore the sample must first be heated to 900–950°C, and then held at this temperature for 20–30 min. Then any sulphur that has appeared on the surface can be removed by ion etching. In subsequent cycles the sample should not be heated above 400–500°C [16]. In all the cases it is reasonable to monitor the surface layer composition by AES. The cleaning cycle may be completed either by annealing or by ion bombardment. The latter provides cleaner surface, but inert gas implantation (of Ar in particular) modifies surface properties causing passivation and producing effects on the kinetics of the initial oxidation stages [56–58].

References

1. Benninghoven, A. and Loebach, E. (1972) Analysis of monomolecular layers of solids by the static method of secondary ion mass spectroscopy (SIMS). *J. Radioanal. Chem.* **12**, 95–100.
2. Fogel' Ya. M. (1976) To the question of choice of current density magnitude for primary ion beam at the study of processes at solid surfaces using SIIE methods. *Zh. Tekhn. Fiz.* **46**, 1767 (in Russian).
3. Benninghoven, A. and Müller, A. (1972) Investigation of the surface oxidation of metals in the submono- and monolayer oxidation range with the static method of secondary ion mass spectrometry. *Thin Solid Films*, **12**, 439–442.
4. Fogel', Ya. M. (1967) Secondary ion emission. *Usp. Fiz. Nauk*, **91**, 75–112 (in Russian).
5. Shvachko, V. I., Nadykto, B. T., Fogel', Ya. M. *et al.* (1965) The use of SIIE method for the study of oxygen interaction with Nb surface. *FTT*, **7**, 1944–1951 (in Russian).
6. Kolot, V. Ya., Rybalko, V. F., Fogel', Ya. M., and Tykhinski, G. F. (1967) On the possibility of SIIE use for the study of corrosion processes. *Zashch. metall.* **3**, 723–729 (in Russian).
7. Rybalko, V. F., Kolot, V. Ya., and Fogel', Ya. M. (1968) Tungsten oxidation at room temperature. *FTT*, **10**, 3176–77 (in Russian).
8. Rybalko, V. C., Kolot, V. Ya., and Fogel', Ya. M. (1969) Investigation of tungsten oxidation by SIIE method (in Russian). *Izv. Akad. Nauk SSSR, Ser. Fiz.* **35**, 836–839.
9. Rybalko, V. F., Kolot, V. Ya., and Fogel', Ya. M. (1969) Study of oxygen adsorption on W by SIIE method. *Zh. Tekh. Fiz.* **39**, 1717–1719 (in Russian).
10. Fogel', Ya. M. (1972) Ion–ion emission — a new tool for mass spectrometric investigation of processes on the surface and in the bulk of solids. *Int. J. Mass Spectrom. Ion Phys.* **9**, 109–125.
11. Kolot, V. Ya., Tatus', V. I., Rybalko, V. F., and Fogel', Ya. M. (1970) Study of oxygen adsorption on Mo surface by SIIE method. *Zh. Tekh. Fiz*, **40**, 2469–2471 (in Russian).
12. Werner, H. W. (1975) The use of secondary ion mass spectrometry in surface analysis. *Surf. Sci.* **47**, 301–323.
13. Benninghoven, A. and Wiedmann, L. (1974) Investigation of surface reactions by the static method of secondary ion mass spectrometry. IV. The oxidation of magnesium, strontium and barium in the monolayer range. *Surf. Sci.* **41**, 483–492.
14. Benninghoven, A. (1975) Developments in secondary ion mass spectrometry and applications to surface studies. *Surf. Sci.* **53**, 596–625.
15. Benninghoven, A. (1976) Neue Methoden zur Untersuchung von Festkörperoberflächen. *Phys. Bohemaslov.* **32**, 298–308.
16. Blasek, G. and Weihert, M. (1979) A study of the initial oxidation of Cr–Ni steel by SIMS. *Surf. Sci.* **82**, 215–227.
17. Kolot, V. Ya., Tatus, V. I., Rybalko, V. Ya., and Fogel', Ya. M. (1971) Study of Mo surface oxide composition by SIIE method. *Izv. Akad. Nauk SSSR, Ser. Fiz.* **35**, 255–260 (in Russian).
18. Fogel', Ya. M., Nadykto, B. T., Shvachko, V. I., and Rybalko, V. F. (1964) Study of oxygen adsorption on an Ag surface using SIIE methods. *Zh. Fiz. Khim.* **38m**, 2397–2402 (in Russian).
19. Fogel', Ya. M., Nadykto, B. T., Rybalko, V. F. *et al.* (1962) On the possibility of using SIIE phenomenon for the study of heterogenous catalitic reactions. *Dokl. Akad. Nauk SSSR* **147**, 414–417 (in Russian).

20. Fogel', Ya. M., Nadykto, B. T., Rybalko, V. F. *et al.* Study of catalytic reaction of ammonium oxidation on Pt using SIIE methods. *Kinet. Katal.* **5**, 496–504 (in Russian).
21. Fogel', Ya. M., Nadykto, B. T., Shvachko, V. I. *et al.* Study of catalytic reaction of ammonium oxidation on Pt using SIIE methods. *Dokl. Akad. Nauk SSSR*, **155**, 171–174 (in Russian).
22. Shvachko, V. I., Fogel', Ya. M. (1966) Study of reaction of ammonium dissociation on Fe using SIIE method. *Kinet. Katal.* **7**, 722–726 (in Russian).
23. Shvachko, V. I., Fogel', Ya. M., and Kolot, V. Ya. (1967) Study of reaction of ammonium synthesis on Fe using SIIE methods. *Dokl. Akad. Nauk SSSR, Ser. Khim.* **172**, 1353–1356 (in Russian).
24. Jede, R., Mauske, E., An, L. D. *et al.* (1984) Silver catalyst for partial oxidation of methanol. Reaction path and catalyst poisoning by iron. A combined SIMS, TDMS, AES, XPS, and ISS study. In: *Secondary ion mass spectrometry, SIMS IV*, Springer, Berlin, pp. 234–237.
25. Shimada, H., Kobayashi, J., Kurita, M. *et al.* (1984) Behaviour of inorganic materials on catalysts used for coal liquefaction. In: *Secondary ion mass spectrometry, SIMS IV*, Springer, Berlin, p. 238.
26. Kybashevski, O. and Gopkins, B. E. (1965) Oxidation of metals and alloys. *Metallurgija* Moscow (in Russian).
27. Rybalko, V. F., Kolot, V. Ya, Fogel', Ya. M. (1969) Oxygen pressure effect on tungsten oxidation process. *FTT.* **11**, 1404–1406 (in Russian).
28. Kolot, V. Ya. Tatus', V. I., Vodolazhchenko, V. V. *et al.* (1972) On the transition of oxide film on Mo surface from two-dimensional phase into three-dimensional one (in Russian). *Zh. Tekh. Fiz.* **42**, 1486–1490.
29. Kolot, V. Ya., Tatus', V. I., Rybalko, V. F. *et al.* About processes which determine the composition of a two-dimensional oxide film on an Mo surface (in Russian). *Zh. Tekh. Fiz.* **42**, 2416–2421.
30. Buhl, R. and Preisinger, A. (1975) Crystal structures and their secondary ion mass spectra. *Surf. Sci.* **47**, 344–357.
31. Stumpe, E. and Benninghoven, A. (1974) Surface oxidation studies of iron using the static method of secondary ion mass spectrometry (SIMS). *Phys. Status Solidi* **21**, 479–486.
32. Dawson, P. H. (1976) The oxidation of Al studied by SIMS at low energies. *Surf. Sci.* **57**, 229–240.
33. Gettings, M., Coad, J. P. (1975) A preliminary study of pure metal surfaces using Auger electron spectroscopy (AES), X-ray photo-electron spectroscopy (XPS) and secondary ion mass spectrometry (SIMS). *Surf. Sci.* **53**, 636–648.
34. Benninghoven, A., Bispinck, H., Ganschow, P., and Wiedmann, L. (1977) Quasi-simultaneous SIMS–AES–XPS investigation of the oxidation of Ti in the monolayer range. *Appl. Phys. Lett.* **31**, 341–343.
35. Benninghoven, A., Müller, K. H., Plog, C. *et al.* (1977) SIMS, EID and flash-filament investigation of O_2, H_2 (O_2 + H_2) and H_2O interaction with vanadium. *Surf. Sci.* **63**, 403–416.
36. Benninghoven, A., Müller, K. H., Schemmer, M., and Beckmann, P. (1978) SIMS and flash desorption studies of Ni–O interaction. *Appl. Phys.* **16**, 367–373.
37. Benninghoven, A., Ganschow, O., and Widemann, L. Quasi-simultaneous SIMS, AES, and XPS investigation of the oxidation of Mo, Ti, and Co in the monolayer range. *J. Vacuum Sci. Technol.* **15**, 506–509.
38. Bispinck, H., Ganschow, O., Wiedmann, L., and Benninghoven, A. (1979) Combined SIMS, AES and XPS investigations of Ta oxide layers. *Appl. Phys.* **18**, 113–117.
39. Abramenkov, A. D., Slezov, V. V., Tanatarov, L. V., and Fogel', Ja. M. (1970) Study of Cu atoms diffusion over Mo surface using SIIE method. *FTT*, **12**, 10, 2929–2933 (in Russian).
40. Werner, H. W. and Boudewijn, P. R. (1984) A comparison of SIMS with other techniques based on ion-beam solid interaction. *Vacuum*, **34**, 83–101.
41. Nemoshkalenko, V. V. and Aleshin, V. G. (1975) Electron spectroscopy of crystals. Naukova dumka, Kiev (in Russian).
42. Morabito, J. M. (1975) A comparison of Auger electron spectroscopy (AES) and secondary ion mass spectrometry (SIMS). NBS Special Publication No. 427. *Secondary ion mass spectrometry*, pp. 33–61.
43. Petrov, N. N. and Abrojan, I. A. (1977) *Diagnostics of the surface using ion beams.* Izd-vo LGU, Leningrad (in Russian).
44. Shul'man, A. R. and Fridrikhov, S. A. (1977) *Secondary emission methods for the study of solids.* Nauka, Moscow (in Russian).
45. Zandern, A. M. (Ed.) (1979) *Methods of surface analysis.* Mir, Moscow (in Russian).

46. Fuller, D., Colligon, J. S., and Williams, J. S. (1976) The application of correlated SIMS and RBS techniques to the measurement of ion implanted range profiles. *Surf. Sci.* **54**, 647–658.
47. MacDonald, R. J., Martin, P. J. (1977) Quantitative surface analysis using ion-induced secondary ion and photon emission. *Surf. Sci.* **66**, 423–435.
48. Martin, P. J. and MacDonald, R. J. (1977) The influence of single crystal structure on photon and secondary ion emission from Ar^+ ion bombarded Al. *Radiat. Eff.* **32**, 177–185.
49. Cherepin, V. T. and Vasil'ev, M. A. (1982) Methods and instruments for the analysis of surface of materials. Naukova Dumka, Kiev (in Russian).
50. Colligon, J. S. (1974) Surface compositional analysis using low energy ion bombardment induced emission processes. *Vacuum*, **24**, 373–388.
51. Ganschow, O. (1984) SIMS combined with other methods of surface analysis. In: *Secondary ion mass spectrometry, SIMS IV*, Springer, Berlin, pp. 213–220.
52. Wiedmann, L., Ganschow, O., and Benninghoven, A. (1978) Contamination of clean metal surfaces associated with electron bombardment in conventional AES analysis. *J. Electron Spectrosc. Relat. Phenom.* **13**, 243–246.
53. Estel, J., Hoinkes, H., Kaarmann, H. *et al.* (1976) On the problem of water adsorption on alkali halide cleavage planes, investigated by secondary ion mass spectrometry. *Surf. Sci.* **54**, 393–418.
54. Kodama, Y., Sumitomo, S., Kato, I. *et al.* Combined spectrometer with the techniques of SIMS, ISS, AES and XPS. In: *Secondary ion mass spectrometry, SIMS IV*, Springer, Berlin, pp. 252–254.
55. Cherepin, V. T., Kosjachkov, A. A., Dubinsky, I. N., and Is'janov, V. E. (1984) Spectrometric complex with angular resolution for the study of solid surfaces. Preprint IMF.11,84, Inst. Metal Phys., Kiev, pp. 1–37.
56. Cherepin, V. T., Ivaschenko, Yu. N., Vasil'ev, M. A. (1973) To the effect of ion bombardment on the corrosion resistance of Fe–C alloys. *Dokl. Akad. Nauk SSSR*, **210**, 821 (in Russian).
57. Vasil'ev, M. A., Kosjachkov, A. A., and Cherepin, V. T. (1976) The effect of Ar and He ion irradiation dose on iron oxidation. *Dokl. Akad. Nauk Ukr. SSR*, Ser. A., 267–269 (in Russian).
58. Cherepin, V. T., Kosjachkov, A. A., and Vasil'ev, M. A. (1976) Effect of initial ion bombardment and oxidation on emission of secondary ions. *Surf. Sci.* **58**, 609–612.

Subject index

T - #0255 - 101024 - C0 - 248/165/8 [10] - CB - 9789067640787 - Gloss Lamination